ITINERARY

OCTAVIO PAZ

———

ITINERARY

An Intellectual Journey

TRANSLATED BY JASON WILSON

A Harvest Book
Harcourt, Inc.
SAN DIEGO   NEW YORK   LONDON

Requests for permission to make copies of any part
of the work should be mailed to the following address:
Permissions Department, Harcourt, Inc.,
6277 Sea Harbor Drive, Orlando, Florida 32887-6777.

www.harcourt.com

This is a translation of *Itinerario*.

Acknowledgment to the *Times Literary Supplement*, where
"Imaginary Gardens: A Memoir" first appeared in July, 1989.

Library of Congress Cataloging-in-Publication Data
Paz, Octavio, 1914–
[Itinerario. English]
Itinerary: an intellectual journey/Octavio Paz;
translated by Jason Wilson.
p. cm.
ISBN 0-15-100562-1
ISBN 0-15-601071-2 (pbk)
1. Mexico—Politics and government.
2. Latin America. 3. Europe. 4. Paz, Octavio, 1914–
I. Wilson, Jason, 1944– II. Title.
F1226 .P3213 2000
972.08'3—dc21 00-037022

Printed in the United States of America
First Harvest edition 2001

A C E G I J H F D B

Epitaph on a nowhere stone

Mixcoac was my village: three syllables of night,
a shadow veil upon a solar face.
Then came Our Lady, mother, Storm of dust.
She came, devoured the village. I traveled the world.
Words became my dwelling place, the air my tomb.

Translated after the Spanish by Anthony Rudolf

CONTENTS

ITINERARY

Foreword

Of the many walks with Octavio that I recall—in Mexico City, Paris, Spoleto, Cambridge, New York—I remember in particular a cool, bright spring afternoon on Fifth Avenue: Manhattan possessed that quality which Matisse especially admired here—the crystalline. The cold air seemed to endow everything with a sharpness of angle, an impinging presence. Suddenly Octavio said, as if the reflection had all at once surprised him, "You know, they made something beautiful by accident here." I suppose he was thinking of the layout of the original grid pattern—in itself an abstraction with no inevitable relation to beauty—that determined the shape of New York before New York had really arrived here, so that when it did arrive there was this underlying form that marvellously chimed with the tall buildings, the light and air. If, on that spring afternoon, the city had become endowed with speech, what we should have heard was "the other voice"— the voice that somehow exceeded the accident of which it was part, that reconciled and transcended the disparate elements in a sort of aria of light and space. That aria would have gone beyond and yet contained the individual separatenesses of

1

street-corner, façade, side-street, doors, windows, of people walking like us.

The union of the seemingly arbitrary—the grid, the abstract pattern—with the busy particulars of daily life, struck me again when Octavio and I wrote together a sequence of poems called *Airborn*. The form we chose was the sonnet with its traditional fourteen lines, ten syllables per line in English and eleven in Spanish (the sequence was to be in two languages). So this preconceived form, like New York's grid pattern, was there to support and to be transformed by all the elements that would flow through it. This transformation of the given (if our efforts were to be any good) must be the work of the imagination, and the imagination must bring to birth something beyond the merely given and beyond our mere daily selves: it must make us and our readers hear the other voice. Of that voice Octavio had already written: "The 'other voice' is my voice. Our being already contains that other wish to be..." And he concludes: "Inspiration is that strange voice that takes us out of ourselves to be everything that we are..."

So the work of the poetic imagination is to subsume and to integrate all that is seemingly arbitrary in poetic form—those pre-existing fourteen lines, those ten syllables to be counted out on the fingers (though in Spanish you need an extra finger).

If, in poetic composition, the other voice is the enemy or rather the vivifier of what is simply given, in history it is the voice always opposing historical determinism, refusing the staleness of fate and the dead-ends that our lives can too easily be led into. Octavio puts it this way: "Every poem is an attempt to reconcile history and poetry for the benefit of poetry," so that, "Poetry is the other voice. Not the voice of history or of anti-history, but the voice which, in history, is always saying something different."

When the other voice makes itself heard, it tends to leave our attempts to pontificate on "the future of poetry" looking rather foolish. Where would you look or listen during the *fin de siècle*—that of the nineteenth century and also our own—to find that voice? One of the places to seek it out last century would have been the unlikely city of Alexandria in Egypt, where two very different poets, Cavafy and Ungaretti, were both born. Amid the leftovers of Oscar Wilde and late Swinburne, who would have thought of listening for that other voice in "a half-savage country" (Pound's phrase), the United States of America? In other words, the arrival of Pound and Eliot and their particular poetries were wholly unforeseen. Similarly, who would have supposed that (only yesterday, so to speak) in a decaying house in Mixcoac, just outside Mexico City, the other voice was preparing to make itself heard in the person of Octavio Paz, born out of and yet transcending the complex fate of being a Mexican and of seeking an identity for oneself and one's country.

"Inspiration," as I have already quoted from Octavio, "is that strange voice that takes us out of ourselves to be everything that we are..." In our age of conceptual art and other gimmicks, we hesitate even to talk about inspiration and yet, in one form or another, it is this power, not disdained by Paz, that unites very different kinds of poets and poetry, and can link together in a more vivid consciousness the entire human race. Inspiration extends from the Bible to the poetry of the American Indians; it overrides distinctions between classicism and romanticism; it can inform the "disciplined kind of dreaming"

T. S. Eliot attributes to Dante, and the (apparently) undisciplined overflow of the unconscious the surrealists felt themselves to be tapping. Basically, the address to the Muse from Homer to Milton and the waiting on a vision—frequently that of some animal—by the American Indians have this in common: a respect for the unpremeditated, for that which lies beyond merely conscious purposes and merely voluntary intentions.

In the person of Octavio Paz, I salute a poet of the tradition of inspiration and of the tradition of the new. The reception of "the other voice" has to be prepared for, and Octavio's unmistakably twentieth-century poetry, a poetry appearing in the aftermath of Eliot's *The Waste Land*, owes much of its accuracy and tonality to his knowledge of earlier poetries—to medieval and seventeenth-century Spanish, to nineteenth-century French (Mallarmé, in particular), to the Nahuatl poetry of his native land. In Paz we have a clear instance of the way the other voice is nurtured by the past and by past forms—even that most traditional of forms, the sonnet. So a poet has this in common with a city—out of the given both create the unforeseen, out of the apparently fated they release new possibilities for life.

CHARLES TOMLINSON

Notice

This book [was originally] made up of two essays. The longest, *Itinerario,* is somewhat autobiographical for it is the story of the evolution of my political ideas. An intellectual biography but also a sentimental and even passionate one: what I thought and think about my time is inseparable from what I felt and feel. *Itinerario* is the story and description of a journey through time, from one point to another, from my youth to my present moment. The line that traces this plan is neither straight nor circular but a spiral that turns back ceaselessly and ceaselessly distances itself from the point of departure. What we are living today brings me close to what I lived seventy years back and, simultaneously, irremediably and definitively distances me. Strange lesson: there is no turning back but there is no point of arrival. We are in transit.

The other text refers to circumstances that drew me to write, more than forty years ago, *The Labyrinth of Solitude.*[1] It is also biographic and refers to my changing relation with my country, its history and its present. Again a meditation and a confession. Overlapping between the texts was inevitable. Where does Mexico end and the world begin? How to distinguish between the past and the present, between what was, what is, and what is going on still in the living tissue of the current situation?

Some may be surprised that after publishing a few months back a book on love, *The Double Flame*,[2] I am now delivering to the public another one whose topic is essentially political. The surprise will vanish as soon as it is noted that love and politics are the two extremes of human relationships: what is public and what is private, the town square and the bedroom, the group and the couple. Love and politics are two poles linked by an arch: the person. The destiny of a person in a political society is reflected in a lovers' relationship, and vice versa. The story of Romeo and Juliet is unthinkable if we suppress the patrician quarrels in the Italian cities of the Renaissance and the same goes for Lara and Zhivago outside the context of the Bolshevik revolution and the civil war. Everything is linked.

OCTAVIO PAZ

How and why I wrote
The Labyrinth of Solitude

Many times have I been asked this question: Why, what for, and for whom did I write *The Labyrinth of Solitude*? There are many answers. The simplest and most direct lies in my infancy. Three moments in my childhood marked me forever and everything that I have written about my country has been no more, perhaps, than an answer to those experiences of childhood vulnerability. A tirelessly repeated answer and, each time, different. The first experience is also my first memory. How old was I? I don't know, maybe three or four years. I remember vividly the place: a small, square room in a grand old house in Mixcoac.[3] My father "had gone to the revolution" as was said then, and my mother and I took refuge with my grandfather, Ireneo Paz, family patriarch. The turmoil of those years had forced him to leave the city and move to his country house in Mixcoac. I lived and grew up in that village, but not always in the same house, apart from a short stay in Los Angeles. I left it just after reaching my twenty-third year. The house still stands, and is today a convent for nuns. Not long ago I paid a visit and could hardly recognize it: the nuns had turned the bedrooms and garden into cells and the terrace into a chapel. It doesn't matter: the image stays with me as do the sensations of wonder and vulnerability.

I see myself—or better still, I see a blurred figure, a child-like bulk lost in a huge round sofa of threadbare silk, placed right in the middle of the room. In a fixed way, light falls from a high window. It must be five in the afternoon for the light is not intense. The walls papered in a faded yellow with drawings of garlands, stalks, flowers, fruits: emblems of boredom. All vivid, too vivid; all alien, closed in on itself. A door gives on to the dining room, another to a drawing room, and the third, at the side and with stained glass, on to the terrace. The three are open. The room was used for breakfast. Drone of voices, laughter, clatter of dishes. It is a holiday, celebrating a saint's day or a birthday. My older cousins rush out to the terrace. There is a coming and going of people who pass by the bulk without stopping. The bulk cries. For centuries he has been crying and nobody hears. He is the only one to hear his wail. He is lost in a world that is both familiar and remote, intimate and indifferent. It is not a hostile world: it is a strange world, although familiar and everyday, like the garlands on the impassive wallpaper, like the laughter from the dining room. Interminable moment: hearing myself cry amidst universal deafness...I do not remember more. Obviously my mother calmed me down: woman is the door that reconciles us with the world. But the sensation has not been wiped out and never will. It is not a wound, but a hollow. When I think of myself I touch it; when I feel myself I feel it. Alien always and always present, it never leaves me, a dumb, invisible, bodiless presence, constant witness to my life. It does not talk to me but I, at times, hear what its silence tells me: that afternoon you began to be yourself—when you discovered me you discovered your absence, your hollow; you discovered yourself. You now know: you are lack and quest.

The ups and downs of the civil war led my father to the United States. He settled in Los Angeles, where there was a large colony of political exiles. A little later my mother and I followed him there. Soon after our arrival my parents decided that I should go to the neighborhood kindergarten. I was six years old and didn't speak a word of English. I vaguely recall the first day of class; the school with its American flag, the empty room, the desks, the hard benches, and how embarrassed I was by my classmates' noisy curiosity and by the affable smile of the young teacher, who struggled to placate them. It was an Anglo-American school and only two of its pupils were Mexican, although born in Los Angeles. Terrified by my inability to understand what they said to me, I took refuge in silence. After an eternity there was a break and lunch. Sitting down at the table I panicked when I realized that I did not have a spoon; I opted not to say anything and refused to eat. One of the teachers, seeing my untouched plate, asked me why in sign language. I mumbled "spoon," pointing to a classmate's one. Someone repeated aloud, "Spoon!" Guffaws and hubbub: "Spoon! Spoon!" Then they started imitating my poor pronunciation, giggling in chorus. The person in charge silenced them but on the way out, in the sandy playground, I was hemmed in by their yells. Some came up close and shouted the dreaded word *Spoon!* into my face, as if spitting at me. One of them shoved me, I tried to hit back and suddenly found myself in the middle of a ring, facing my aggressor with fists raised like a boxer as he challenged me shouting, "Spoon!" We laid into each other until pulled apart by a teacher. After school we were reprimanded. I did not understand an iota of

9

the ticking off, but I returned home with my shirt ripped and with three scratches and a swollen eye. I did not go back to the school for a fortnight; then, little by little, things became normal; they forgot the word *spoon* and I learnt how to pronounce it properly.

The political outlook changed in Mexico and we went back to Mixcoac. In keeping with family tradition, my parents sent me to a French school of the Salesian order. Although I spoke English, I had not forgotten my Spanish. However, my classmates quickly decided that I was a foreigner; a gringo, a Frenchy, a dago, it was all the same to them. Aware that I had just arrived back from the United States and given my complexion—brown skin and hair, blue eyes—their attitude was easily explained, but not completely: my family was known in Mixcoac from the beginning of the century and my father had stood for the municipality. Once again I suffered laughter and giggles, nicknames and fights, sometimes at the school football pitch, sometimes in an alley near the church. I often got home with a black eye, bruised lips, and a scratched face. At home they were worried, but wisely did not interfere; things calmed down bit by bit, on their own. That's how it was, although the grudge persisted: the slightest pretext was sufficient for the inevitable insults to burst out again.

My experiences in Los Angeles and in Mexico weighed down on me for many years. Sometimes I felt guilty—we are often accomplices of our persecutors—and would say to myself: yes, I am neither from here nor from there. Then, where am I from? I felt Mexican—my surname Paz first appeared in Mexico in the sixteenth century, just after the Conquest—but *they* did not let me be Mexican. I once accompanied my father to a friend he rightly admired: Antonio Díaz Soto y Gama,[4] the old, Quixotic, Zapatista revolutionary. He was in his office with several friends and, when he saw me, he bellowed to my father: "Goodness, you never told me you had a visigothic son!" Everybody laughed at his joke but I took it as a condemnation.

Although the background to the three experiences was similar—a feeling of separation—each one was different. The first is universal and common to all men and women. Theologians, philosophers, and psychologists have written many pages on this matter; it has been a favorite subject for great poets and novelists have ceaselessly explored its bypaths. We are children of Adam, the first exile. The experience faces us with universal indifference, that of the cosmos and of our fellows; at the same time, it is the source of our thirst for a totality and participation that we all yearn for from birth. The second and third are historical and the consequence of that reality that is basic to political organization: the human group, the community. Nothing more natural than that a Mexican child should feel strange in a North American school, but it is awful that the other children should insult him and harm him merely because he is a foreigner. Awful, natural, and as old as human societies. It was not by chance that the wary Athenians invented the punishment of ostracism for those under suspicion. And the foreigner is always under suspicion. The third experience can be assigned to this last category: I was not, clearly, a foreigner but, thanks to my looks and other moral and physical circumstances, I was suspicious. Thus my classmates condemned me to exile, not abroad but at home.

I am not the first, of course, to suffer this sentence. Nor will I be the last. However, although it is a fact that belongs to all times and all places, some people are more prone than others to find suspicious people everywhere... and condemn them to ostracism within or outside the city. I already mentioned the

Athenians. Another people eaten up by suspicion are the Mexicans. The psychological roots of this propensity are suspicious in themselves. Whether it is in a Greek from the fifth century before Christ or a Mexican from the twentieth century, suspicion is the expression of a feeling of insecurity. During crises and social upheavals, mistrust flourishes; Robespierre, called by some Incorruptible and by others Tyrant, embodied suspicion dressed up as revolutionary vigilance. In the twentieth century Bolsheviks repeated and exaggerated this model; contrariwise, one of Julius Caesar's traits that most surprised the ancients was his confidence. Some admired him for it, others rebuked him for it: a confident dictator is a political scandal and a moral contradiction. Suspicion is related to malice and both serve envy. If the public circumstances are right, all these evil passions become accomplices to inquisitions and repressions. Betrayal and calumny are procurers for tyrants.

In Mexico, suspicion and mistrust are collective diseases. In my youth I witnessed the harassment suffered by the writers named the Contemporáneos, after the magazine they published.[5] They were accused of being foreign-lovers, cosmopolitans, Francophiles, in fact, of not being Mexicans. They were a sickly, foreign body embedded in our literature: they had to be ejected from the Republic of Letters. (At the time that I was editing *Plural* with a group of friends, a young Marxist philosopher also demanded that we be expelled from "political discourse.")[6] Ideological and sexual orthodoxy are always linked to xenophobia: the Contemporáneos were accused of being aesthetic reactionaries and branded as queers. Today, young writers revere their memory and write ardent essays on them. Few remember that whilst they were alive they were seen as suspicious and sentenced to exile in their own land. Years later I ceased being a witness to the malignancies

12

of suspicion and was made the object of similar campaigns, although somewhat more fierce: political passions were added to the malice of old.

It is not strange that, due to all this, I have been intrigued by Mexican suspicion since my adolescence. It seemed to me to be the consequence of an inner conflict. After meditating on its nature, I found that it was the result of a historical wound buried in the depths of the past, rather than a psychological enigma. Suspicion, always awake, ensures that nobody discovers the corpse and digs it up. That is its psychological and political function. Now, if the root of the conflict is historical, only history can clear up the enigma. The word *history* suggests first of all a process, and when you say process you mean quest, usually an unconscious one. Process is quest because it is movement and all movement is a "going towards."[7] But, towards what? It is not easy to answer this question: the supposed ends of history have been vanishing one after the other. Perhaps history has no finalities, no end. The sense of history is ourselves, we who make it and by making it unmake ourselves. History and its meanings will end when humans die out. However, although it is impossible to detect ends in history, it is not hard to affirm the reality of the historical process and its effects. Suspicion is one of them. What I have called the *quest* is the attempt to resolve this conflict that suspicion perpetuates.

Without clearly understanding what I was doing, moved by an intuition and stung by the memory of my three experiences, I wanted to rip the veil apart and *see*. My act was an interrogation that linked me to history's unconscious process, that is, to the quest which forms the basis of historical movement. My interrogation inserted me in the quest, made me part of it: thus what began as a private meditation turned into a thinking about Mexican history. This thinking took the shape of a question not only about its origins—where and when did the conflict start?—but also on the meaning of the quest that is Mexican history (and everybody else's history). Obviously, nobody knows with any certainty what we are seeking but we all know we are seekers. Is there anything else we should know? Whilst I was thinking, my three childhood experiences revealed their dual nature to me: they were private and collective, mine and everybody's.

For millennia the American continent lived a life apart, ignored by and ignorant of other people and other civilizations. European expansion in the sixteenth century broke this isolation. Real universal history does not start with the great European and Asiatic empires, with Rome and China, but with the explorations of the Spanish and Portuguese. Since then we Mexicans have been a fragment of world history. Better said: we are children of that moment in which the different histories of peoples and civilizations flow into universal history. The Discovery of America initiated the planet's unification. The act that founded us has two faces: the Conquest and the evangelization; our relationship with it is ambiguous and contradictory: the sword and the cross. No less ambiguous is our relationship with Mesoamerican civilization; its spectre inhabits our dreams, but it rests forever in the great cemetery of vanished civilizations. Our birthplace was a battle. The meeting between the Spaniards and the Indians was simultaneously, to use the poet Jáuregui's lively, picturesque image,[8] burial mound and marriage bed.

Perhaps through family influence, the history of Mexico has been my passion since childhood. My grandfather, author of historical novels conforming to nineteenth-century taste, had collected a good number of books on our past. One topic fascinated me above all others: the clash between people and civilizations. The nations of ancient Mexico lived in constant war one against the other but it was only with the arrival of the Spaniards that they really faced the *other*, that is, a civilization different from their own. Later, already in the modern period, we had violent encounters with the United States and with the France of the Second Empire. Although French culture's influence was very strong in the second half of the nineteenth century and first half of the twentieth, the war with France did not have any further political consequences. Nor did it have psychological ones. The opposite happened with Spain and the United States: our relationship with these nations has been polemical and obsessive. Every country has its phantoms: France for the Spaniards, Germany for the French, ours have been Spain and the United States. The phantom of Spain is losing its significance and its political and economic influence has dried up. Its presence is psychological: a genuine phantom, it haunts our memory and kindles our imagination. The United States is a reality but one so vast and powerful that it borders on myth, and for many, on obsession.

The quarrel between hispanists and antihispanists is a chapter of Mexico's intellectual history. Also of its political and sentimental history. The faction of the antihispanists is not homogeneous: some adore Mesoamerican culture and condemn the Conquest as genocide; others, less numerous, descendants of the nineteenth-century liberals, profess an identical scorn for both traditions: the Indian and the Spanish, both obstacles on the path to modernization. I was familiar with this dispute from my childhood. My paternal family was liberal, and also in favor of the Indians: anti-Spanish on

15

two counts. Although my mother was Spanish, she loathed arguments and answered diatribes with a smile. I found her silence sublime, more crushing than a tedious speech. Otherwise, in my grandfather's library there were countless books that contradicted his moderate antihispanism and my father's keener one. Both identified the Spanish past with the ideology of their traditional enemies, the conservatives. Galdós opened my eyes: that tussle was also Spanish.

The anti-Spanishness in those close to me was of a historical and political cast, not literary. Amongst my grandfather's books were our classics. Moreover, he admired the Spanish liberals of the last century. My adolescence and youth coincided with the end of monarchy and the first years of the republic, a period of true splendor in Spanish letters. Reading the great writers and poets of those years ended up by reconciling me with Spain. I felt part of the tradition—not in a passive way, but actively, at times, even polemically. I discovered that the literature written by us Hispanoamericans is the other face of the hispanic tradition. Our literature began by being a tributary of Spanish letters, but is now a powerful river. Cervantes, Quevedo, and Lope would recognize themselves in our authors. The dispute between hispanists and antihispanists seemed to me to be anachronistic and sterile. War in Spain soon silenced this debate for good. At least for me and many like myself. I was a passionate partisan of the Republican cause and went to Spain in 1937 for the first time. In several essays and in some poems I have spoken of my encounter with its people, its landscapes, its stones. I did not discover Spain: I recognized it and I recognized myself.[9]

My experience of North American reality was also, in its own way, a confirmation. In my childhood I had lived in California but the true confrontation began in 1943 and lasted until December 1945. I lived in San Francisco and in New York, I spent a summer in Vermont[10] and two weeks in Washington, I took several jobs, I dealt with all kinds of people, I had financial difficulties, I lived through highs and lows, I voraciously read English and North American poets and, at last, started to write poems free from the rhetoric that stifled the poetry written by young poets of that time in Spain and in Latin America. In a word, I was born again. I had never felt so alive. These were the war years and North Americans were living one of the great moments of their history. In Spain I had known fraternity in the face of death; in the United States friendliness in the face of life. A universal sympathy with roots not in the puritanism which, shackled to purity, is an ethics of separation, but the Romantic pantheism of Emerson and the cosmic effusion of Whitman. In Spain some Spaniards recognized me as one of them; in the United States some North Americans welcomed me like a long lost brother who spoke their language with a strange accent and terrible syntax.

My admiration and sympathy for North Americans had a dark side: it was impossible to close my eyes to the fate of the Mexicans, those born over there and the newly arrived. I thought about the years spent in Los Angeles, in my father's struggle to make a living in exile, in my mother working hard as an ant, but an ant who sang like a cicada.[11] Although we did not suffer the hardships of the majority of Mexican immigrants, not much imagination was needed to understand them and feel deeply for their plight. I recognized myself in the *pachucos* and in their mad rebellion against their present and their past.[12] A rebellion that ended not as an idea but in a gesture. The underdog's option: the aesthetic application of defeat, the revenge of imagination. I came back to the question about myself and my destiny as a Mexican. The same one

I had put to myself in Mexico, reading Ortega y Gasset or talking with Jorge Cuesta on a patio in San Ildefonso.[13] How to answer it? Before abandoning Mexico, a year before, I had written a series of articles for a newspaper in which I dealt with topics that were more or less connected with the question that tormented me.[14] They no longer satisfied me. I had no idea then that these notes and my discoveries in Spain and in the United States were preparations for the writing of *The Labyrinth of Solitude.*

I reached Paris in December 1945. In France the years in the wake of the Second World War were of dearth but great intellectual liveliness. It was a period of great riches, not so much in the domain of literature itself, of poetry and novels, but in ideas and essays. I zealously followed the philosophical and political debates. A burning atmosphere: passion for ideas, intellectual rigor and, at the same time, a marvellous sense of freedom. I soon found friends who shared my intellectual and aesthetic anxieties. In those cosmopolitan circles—made up of French, Greeks, Spaniards, Romanians, Argentines, North Americans—I could breathe freely: I did not belong there, and, yet, I felt that I had found an intellectual homeland. A homeland that did not ask for identity papers. But the question about Mexico did not abandon me. Having made a decision to face up to it, I drew up a plan—I never managed to follow it completely—and started to write. It was the summer of 1949, the city was deserted, and my work in the Mexican embassy, where I held a modest post, had slackened off. Distance helped me; I lived in a world far removed from Mexico and immune to its phantoms. I had Friday afternoons and all Saturdays and Sundays to myself. And the nights. I wrote quickly and fluently, keen to finish as soon as possible, as if a revelation awaited me on the last page.

I was racing against myself. Who or what was I going to meet at the end? I knew the question, not the answer. Writing became a contradictory ceremony, made up of enthusiasm and rage, sympathy and anxiety. As I wrote I settled a score with Mexico; a moment later, my writing turned against me and Mexico took its revenge. Inextricably knotting together passion and lucidity: *I hate and I love.*

I have elsewhere alluded to the defects and gaps in *The Labyrinth of Solitude.* The first are congenital, the natural consequence of my limitations. As for the latter: I tried to remedy them in diverse texts, as the reader of this book will note. The greatest omission was that of New Spain: the pages I dedicated to it are insufficient; I have expanded these in several essays, especially in the first part of my study of Sor Juana Inés de la Cruz.[15] And the pre-Hispanic world? I think that my essays on the ancient art of Mexico are somewhat more than mere aesthetic exercises: they are a vision of Mesoamerican civilization. Having said this, I confess that the central idea of *The Labyrinth of Solitude* still seems valid to me. The book is not an essay on an illusory "philosophy of Mexican man," nor is it a psychological description, nor a portrait. The analysis starts with some characteristic traits but quickly turns into an interpretation of Mexican history and our situation in the modern world. The interpretation seems valid to me, not exclusive, not total. There are other interpretations, and some of them are (or could be) equally valid. They do not exclude mine because none are global or final. Historical understanding is, by its nature, partial, whether it's by Thucydides or by Vico, by Marx or by Toynbee.

All visions of history are a point of view. Naturally, not all points of view are valid. So, why does mine seem valid? Well, because the idea that inspired it—the dual rhythm of solitude and communion, the feeling alone, divided, and the desire to be reunited with others, and with ourselves—can be applied to all people and all societies. Although every individual is unique and every people is different, we all go through the same experiences. Thus it is licit to present Mexican history as a flow of ruptures and unions. The first was the Conquest— first and most decisive: it was a collision of two civilizations and not, as would happen later, within the same civilization. At the same time, the first reunion or reconciliation—answer to the violent rupture of the Conquest—consisted of the conversion of the vanquished to a universal faith, Christianity. Since then, ruptures and reunions have followed each other; it would be pointless enumerating them. No, it is not arbitrary to envision our history as a process ruled by the rhythm—or dialectic—of what is closed and what is open, solitude and communion. Furthermore, it is not hard to heed that this same rhythm rules the histories of other people. I think I am dealing with a universal phenomenon. Our history is but one version of this perpetual separation and union with themselves that has been, and is, life for everybody in all societies.

The process of successive ruptures and reunions can also be seen, to employ an analogy with physics, as a series of explosions. Modern cosmology has familiarized us with the idea of an infinitely concentrated matter which, when it reaches a certain density, explodes and scatters. Historical explosions are similar to a *big bang*: a society locked up in itself is doomed to explode when its elements collide. Contrary to what happens in the cosmos, subject apparently to an endless expansion, the elements scattered in history tend to regroup. These new combinations can be translated, once more, into new historical forms. If the rupture is not resolved as reunion, the system dies out, usually absorbed into a greater system. The history of

Mexico fits the first model and can be seen as a succession of explosions followed by dispersions and reunions. The last explosion, the most powerful, was the Mexican revolution. It shook the whole social fabric and managed, after scattering them, to regroup all Mexicans into a new society.

The Revolution rescued many groups and minorities who had been excluded as much from the society of New Spain as from the republic. I am referring to the peasant communities, and to a lesser extent, to indigenous minorities. Moreover, it managed to create an awareness of national identity that hardly existed before. In the sphere of ideas and beliefs, it succeeded in reconciling modern with ancient Mexico. I emphasize that it was a reconciliation at the emotional and spiritual levels, not at the intellectual one. The Revolution was, above all else, a political and social triumph, but it was more, far more: a radical change in our history. As the word *change* turns out to be ambiguous, let me add that this change was a return. I mean: it was a genuine revolt, a return to the origins. In that sense, the revolutionary movement prolonged, at a psychic level different to religion's, the syncretism of the sixteenth and seventeenth centuries. It prolonged it without anybody having decided it, not the leaders, nor the people; however, everybody was moved by the same dark impulse. The logic of history, or popular instinct? It is not easy to know. What is clear is that Mexico hurled itself on to the path of self-knowledge. In an act of necessary rupture, liberalism negated the new-Hispanic and indigenous traditions. The Revolution initiated the reconciliation with our past, something that seems to me not less but more imperative than all the projects of modernization. In this can be found both its originality and its fertility at the level of feelings, beliefs, letters, and arts.

To understand its unique character, it must be remembered that our revolution owes very little to the revolutionary ideologies of the nineteenth and twentieth centuries. In this sense it was the antithesis of 1857 liberalism. This liberalism was a movement derived from universal ideas of European origin; with these ideas the liberals proposed to change society from its roots. Thus their hostility to both the Spanish and the indigenous traditions. The liberalism of 1857 was a true revolution and its archetypes were the French Revolution and United States War of Independence. Contrary to this, the Mexican Revolution was popular and instinctive. It was not guided by a theory of equality: it was possessed by an egalitarian and communitarian passion. The origins of this passion lie, not in modern ideas, but in the tradition of indigenous communities before the Conquest and in evangelical Christian missions. If one reviews the declarations and speeches of the chiefs and popular leaders, it is surprising, to start with, to find so many references and quotations from primitive Christianity. The most used were the Sermon on the Mount and the Expulsion of the merchants from the Temple.[16] Also, it is notable how stubbornly the peasant movement sustained, as the core of its aspirations, the communitarian traditions of its people. Peasants demanded the *return* of their lands.

Can one talk of a revolutionary ideology? The answer is subtle. At a first level, the revolution passed through different moments, and in each one certain themes and ideas predominated. For example, in the first period political reform and the installation of a genuine democracy seemed to be essential; later, social grievances and egalitarian aspirations were crucial; even later, political stability and economic development; and so on. On top of the changes of ideas in time should be placed differences in space: the movement in the south was primordially agrarian, finding its inspiration in a tradition of struggle for communal land that derived from New Spain and the new-Hispanic past; in the north, the nucleus of the movement was made up of ranchers; in the cities by the middle class. Moreover, during this process, there was armed struggle between leaders and factions. The Revolution was many revolutions.[17]

As for the influence of ideologies from abroad, the most substantial, though not preponderant, were: anarchism, the liberal inheritance, trade-unionism—echoes from Chicago's 1st of May—and, perhaps, a vague but strong dream of social redemption. Most crucial, however, was the egalitarian and communitarian current, double legacy from Mesoamerica and New Spain. Not so much a clearly defined doctrine as a bunch of aspirations and beliefs, a subterranean tradition believed to have disappeared and that was revived in the great revolutionary shake-up. It was not easy for this confused and clear-sighted bunching of aspirations, affronts, hopes and grievances to define itself in a clear reform project. This explains why the Revolution ended in a compromise between the liberal inheritance of 1857, the popular communitarian aspirations, and scraps of other ideologies.

Influences from abroad appeared in a later period, when the triumphant revolutionary factor had already established itself in power, and the popular movement had been transformed into an institutional regime. Inspired by the Soviet example (the *kolkhoz*), Lázaro Cárdenas modified communal ownership of the land.[18] The reform did not free the peasants: it tied them to the state banks and turned them into tools of government policy. Cárdenas also started the statist policy in economic matters, followed by nearly all his successors. One of the consequences of the nationalization policy was the appearance of a powerful bureaucracy embedded in the state. Another factor, perhaps the most decisive, that explains the extraordinary growth of bureaucracy was the creation of a state hegemonic party, in power since 1930. The founder of the party was president Calles; two other presidents, Cárdenas and Alemán, consolidated it through successive reforms. The models for this party were the Fascist Party in Italy and the Communist Party in Russia. However, not once did the Mexican party reveal totalitarian ideological ambitions. It was and is a party *sui generis,* the result of a compromise between authentic democracy and revolutionary dictatorship. The compromise prevented civil war between the revolutionary factions and assured the necessary stability for social and economic development.

If one studies the Mexican Revolution from the perspective that I have sketched, one immediately notices that the second period, the so-called institutional, does not only present radical differences with the first but cannot strictly be called revolutionary. The protagonists of the second period have been and are professional politicians; they belong to the middle class, and nearly all have been to university. The ruling élite is a strange but not uncommon amalgam of politicians and technocrats. Thus, in a strict sense, the Mexican Revolution should be viewed as a movement that begins in 1910 and dies out in 1930, with the founding of the Revolutionary Mexican Party. Those twenty years were not only rich in dramatic and at times atrocious military episodes, but fertile in ideas and predictions. Much was destroyed, as much or more than during our terrible Independence war, but also plenty was created. What differentiates this period, above and first of all, is popular participation: people really carried out the revolution, not a group of theorists and professionals as elsewhere. Because of all this it is not risky to state that our movement better fits the old notion of *revolt* than the modern concept of *revolution*. In other writings I have dedicated some speculations to the differences between revolt and revolution.[19] I cannot dwell here on this topic and limit myself to underlining that the notion of *revolt* can be naturally inserted in the image of historical explosion: a rupture that is, also, an attempt at reuniting the scattered elements, solitude and communion.

Between 1930 and 1940, as much in Europe as in America, the majority of writers who were then young felt an immense sympathy for the Russian Revolution and communism. In our attitudes we mixed together decent feelings, a justified indignation faced with injustices around us, and ignorance. Had I written *The Labyrinth of Solitude* in 1937 I would doubtlessly have affirmed that the meaning of the Mexican Revolutionary explosion—what I have called its *quest*—would have ended by adopting communism. The communist society was going to solve the dual Mexican conflict, the inner and the outer: communion with ourselves and with the world. But the period that runs from 1930 to 1945 was not solely one of faith and noisy support but one of criticism, revelations, and disappointments. My doubts began in 1939: in 1949 I discovered the existence of concentration camps in the Soviet Union and from then on it did not seem so clear that communism was the remedy for suffering in the world and in Mexico. My doubts turned into criticisms, as can be seen in the second edition of my book (1959) and in other writings. I saw communism as a bureaucratic regime, petrified into castes, and I saw the Bolsheviks who had decreed, under penalty of death, an "obligatory communion," fall one after the other in those public ceremonies of expiation that were Stalin's purges. I understood that authoritarian socialism was not the *resolution* of the Mexican Revolution, in the historical and musical sense of the word: a step from discord to harmony. My criticisms unleashed a bilious eruption of vituperations in many virtuous Mexican and Spanish-American souls. The wave of hate and silt lasted many years; some of the spray is still fresh.

At the same time as the revolutionary solution was closing, further historical perspectives were opening. It is obvious that our country's and the world's new situation demanded a radical change of direction. A marginalized nation, we had been history's butt: the second half of the twentieth century—marked by colonial independence and by agitation, revolt, and revolution in the peripheral countries—faced us with other realities. I wrote in the last pages of my book: "We have ceased to be the butts and are beginning to be subjects of historical changes." And added: "the Mexican Revolution flows into universal history... there nakedness and neglect await us." Indeed, the collapse of ideas and beliefs, both traditional and revolutionary ones, was universal: "We are at last alone facing the future, like everybody... We are now contemporaries with all people..." The solitary's luck: *testis unus, testis nullus*.[20] Nobody listened: Mexico did not change direction, the governments did not tackle reform but continued their routines and merely survived, whilst intellectuals gripped on to more and more limited and caricatural versions of Marxism. Some interpreted one of my opinions—"we are contemporaries with all people"—as confirmation of our country's maturity: at last we had caught up with the other nations. Curious notion of history as a race: against whom and towards where? No, history is an intersection between time and place. History, said T. S. Eliot, is here and now.[21]

I chose a path that, once again, allowed me to be called into question by the majority of Latin-American writers, at the time still dazzled by the will-o'-the-wisp of "real socialism."

With a few others I argued that only the installation of an authentic democracy, with a legitimate regime and guarantees for the individual and for minorities, could ensure that Mexico did not sink into the ocean of universal history, infested with Leviathans. Modernization, a word not yet in fashion, was both our condemnation and our salvation. Condemnation because modern society is a long way from being exemplary: many of its manifestations—advertising, the cult of money, abysmal inequalities, ferocious egoism, the uniformity of tastes, opinions, minds—are a compendium of horrors and stupidities. Salvation because only a radical transformation, through genuine democracy and the dismantling of the patrimonialism inherited from the viceroyalty (reflecting in turn the European absolutism of the seventeenth and eighteenth centuries), could give us the confidence and strength to face a chaotic and pitiless world. Many of the post-revolutionary institutions, adopted to start with as transitory measures, had already lost their usefulness and reason for being. Others were a frank usurpation of the function usually reserved for the private sector. The unions and other popular associations lived under official tutelage through the monopoly of the government party (a situation that still persists in many ways). To sum up, a system of sly gifts and punishments intended to attract or silence independent opinion. We were not a dictatorship but we were a society under a paternalistic regime that lived between the threat of control and the prize of subsidy. The urgent task was to give the initiative back to society. For all this, although *The Labyrinth of Solitude* was a passionate denunciation of modern society in its two versions, capitalist and totalitarian, it does not end by preaching a return to the past. On the contrary, it underlines that we ought to think it out ourselves and face a future common to everybody.

Universality, modernity, and democracy are today inseparable terms. Each one depends on and demands the presence
of the others. This has been the theme of all that I have written on Mexico since the publication of *The Labyrinth of Solitude*. It has been an acrimonious struggle and has gone on for
too long. A struggle that has tested my patience because of the
abundant blows below the belt, malicious insinuations and
slanderous campaigns. Defending modern democracy, I must
admit, has not been and is not easy. Never once have I forgotten the injustices and disasters of liberal, capitalist societies. The shadow of communism and its jails could have
hidden contemporary reality; its collapse has allowed us to see
capitalist societies in all their desolation: the desert expands
and covers the whole earth. Amongst the ruins of totalitarian
ideology now sprout ancient and ferocious fanaticisms. Present time inspires in me the same horror that I experienced in
my adolescence facing the modern world. *The Waste Land,*
that poem that so impressed me when I discovered it in 1931,
continues to be deeply topical. A moral gangrene corrodes
modern democracies. Are we living the end of modernity?
What awaits us? . . . I halt here: reaching this point, my reflection on Mexico closes and what develops in the following
essay opens. I limit myself to repeating; yes, the children of
Quetzalcóatl and Coatlicue, of Cortés and la Malinche enter
now on their feet into the history of all people, and not
pushed by a stranger.[22] The lesson of the Mexican Revolution
can be distilled into this sentence: we sought ourselves and
found the others.

MEXICO
9 December 1992

Itinerary

Certain historical periods reveal a harmony between customs and ideas. For example, in the twelfth and thirteenth centuries social practices corresponded with beliefs, as these did with ideas. There were great differences between a peasant's faith and a theologian's speculations, but there was no break. The ancient image of the "chain of being" fits medieval society perfectly. The Age of Modernity, since the Renaissance, has been one of rupture; for more than five hundred years we have lived discordantly between ideas and beliefs, philosophy and tradition, science and faith. Modernity means being split in two. The separation began as a collective phenomenon and from the start of the second half of the nineteenth century, as Nietzsche noted before anyone, it became internalized and divided every consciousness. Our time is one of split consciousness, and of being conscious of the split. We are divided selves in a divided society. Discord between customs and ideas was the origin of another characteristic of the Modern Age; the unique trait that distinguishes it from all other periods is the pre-eminence of the word "revolution" from the end of the eighteenth century on. The word and the concept: revolution is the idea embodied in a group and converted as much into a weapon for combat as into a tool to build a new society. Revolution: a theory about change, the act that carries it out and the building of the house of the future. The revolutionary is a type of man that combines the attributes of a philosopher, a strategist and a social architect.

The concept of *revolution* in the triple sense outlined above, was completely unknown by the societies of the past as much in the West as in the East. Those societies, not excluding the primitive ones, always viewed change suspiciously, even with horror; they all venerated an invariable principle whether an archetypal past, a divinity or whatever other concept that meant the superiority of a static being over evolution. Modernity has been unique in overvaluing change. This overvaluation explains, moreover, the emergence of the idea of revolution. The closest to this idea is the founding of a new religion; the advent of a new faith has always been, like a revolution, a rupture and a beginning. But the similarity between these two phenomena does not hide obvious and radical differences. Whatever may have been their upheavals and mishaps, ancient societies knew nothing about revolutionary changes in the strict sense of that word; they knew religious changes. The basis of those changes was quite different from the basis of the revolutionary ones: a divine revelation, not a philosophical theory. Their temporal horizon was also different; not a future but a supernatural beyond. These differences do not annul the similarity, noted before, between religion and revolution: both are answers to the same psychic necessities. In that way the revolutions of the Modern Age have claimed to substitute religions in their dual function: change people and give a meaning to their presence on earth. Now we can see that they were false religions.

The triumph of the revolutionary idea could not close the

breach, open since the Renaissance, between customs and ideas, between belief and theory. The sciences and modern philosophy have grown and developed in an independent manner, at times even hostile to revolutionary thinking. There is no obvious relationship between, for example, Newton's theories and Robespierre's, or between Lenin's and the exact science of the twentieth century. The same goes for philosophy, art, and literature. Neither Balzac, nor Proust, nor Kafka could properly be called revolutionary artists. On the other hand, Dante was not only a Christian poet but his work is inseparable from the medieval mind and philosophy. To summarize, revolution presented itself as a true idea, born of philosophy and science, and this distinguishes it from religion founded on supernatural revelations; at the same time, for the exact sciences and authentic philosophy, revolutionary theories are not and never were science or philosophy. The challenge as much as the final misfortune of the revolutionary idea can probably be traced to this original ambiguity: it was neither a true religion nor a true science. What has it been, then? Generous passion and criminal fanaticism, illumination and darkness. These pages are the witness of a Mexican writer who, like many others of his generation at home and around the world, lived those hopes and disillusions, that frenzy and that disappointment.

FIRST STEPS

1929 saw the start of a Mexico that is now ending. It was the year the National Revolutionary Party was founded and it also saw the birth and failure of a powerful movement of democratic opposition led by an intellectual, José Vasconcelos. The Revolution had been transformed into an institution. The country, bled by twenty years of civil war, was licking its wounds, restoring its strength, and wearily trying to walk. I was fifteen years old, was studying to enter university and had taken part in a student strike that paralyzed the university and stirred the country. The following year I entered San Ildefonso College, an ancient Jesuit seminary converted by the revolutionary governments into the National Preparatory School, the door that led to university faculties. There I met José Bosch, a fellow agitator in the student movement the year before. He was a Catalan, a little older than me. Thanks to him I discovered my first libertarian writers (his father had militated in the Iberian Anarchists Federation). We soon made friends with people who shared our concerns. At San Ildefonso I did not change skin nor soul: those were not years of change but the start of something that still has not ended, a circular search that has been a perpetual beginning again: to discover the reason for those continuous agitations that we call *history*. Years of initiation and apprenticeship, first steps in the world, first time I was led astray, attempts to enter into myself and talk with that stranger I am and will always be for myself.

Youth is a time of loneliness but also of passionate friendships. I had many and was, as is said in Mexico, a great friend of my friends. One of them decided to organize a students' union backing workers and peasants; ostensibly it was dedicated to popular education but it also, and with greater determination, served to spread our vague revolutionary ideas. We used to meet in a tiny room in the college that soon became a center for arguments and debates. It was the seedbed for several, conflicting political futures: some ended up in the official party and held high posts in public administration; a few others, nearly all Catholics, some influenced by Maurras, some by Mussolini, and some by Primo de Rivera, tried unsuccessfully to found fascist parties; the majority turned to the left, and the most daring joined the Communist Youth. The tireless Bosch, faithful to his libertarian ideas, argued with everybody but did not manage to persuade anyone. Slowly he found himself out on a limb. In the end he vanished from our lives as suddenly as he had arrived. He was a foreigner, his papers were out of order, he often fought in student brawls, and the government finally expelled him from the country, despite our protests. I saw him again briefly in 1937 in Barcelona before he was swallowed by the Spanish whirlwind.

Politics was not our sole passion. We were even more attracted by literature, the arts, and philosophy. For me and for a few of my friends, poetry turned, if not into a public religion, at least into an esoteric cult wavering between the catacombs and the conspirators' basement. I found no contradiction between poetry and revolution: they were two facets of the same movement, two wings of the same passion. This belief would link me later to the surrealists. A plural avidity: life and books, street and cell, bars and loneliness in crowds in cinemas. We discovered the city, sex, alcohol, friendship. All these meetings and discoveries were muddled with the images and theories that flowed from our chaotic reading and conversations. Woman was a fixed obsession that constantly changed face and identity: at times she was called Olivia, at others Constanza; she would appear turning a street corner or surge out from the pages of a D. H. Lawrence novel; she was Poetry, Revolution, or the woman next to me in a tram. We read the Marxist catechisms of Bukharin and Plekhanov then the next day we plunged into the electric pages of *The Gay Science* or into the elephantine prose of *The Decline of the West*.[23] Our great supplier of theories and names was the *Revista de Occidente*. The influence of German philosophy was such in our university that in the course on logic our basic text was by Alexander Pfänder, a disciple of Husserl's. Next to phenomenology was psychoanalysis. During those years Freud's works began to be translated and the few bookshops in Mexico City were suddenly flooded by the usual popularizing editions. Many were drowned.

Other magazines were vantage points from where we could first glimpse and then explore the vast and confused territories of literature and art that were always shifting: *Sur, Contemporáneos, Cruz y Raya.*[24] Through these magazines we discovered the modern movements, especially what was happening in France, from Valéry to Gide, the surrealists and the writers in the *Nouvelle Revue Française.* With a mixture of admiration and confusion we read Eliot and Saint-John Perse, Kafka and Faulkner. But none of our admirations tarnished our faith in the October Revolution. That is why one of the writers who most fascinated us was André Malraux, in whose novels we saw the union of modern aesthetics with radical politics. A similar feeling was inspired in us by Thomas Mann's novel *The Magic Mountain*; many of our arguments were naïve parodies of the dialogues between the idealist liberal Settembrini and the Jesuit communist Naptha. I remember that when I met Jorge Cuesta[25] in 1935 he pointed out the disparity between my communist sympathies and my tastes and aesthetic and philosophical ideas. He was right, but the same reproach could have been made about Gide, Breton, and many others, including Walter Benjamin himself. If the French surrealists had declared themselves communists without renouncing their principles and if the Catholic Bergamín[26] announced his adherence to the revolution without giving up his faith, why not forgive our contradictions? They were not ours; they belonged to the age. In the twentieth century the split became an innate condition: we really were divided souls in a divided world. However, some of us managed to transform this psychic crack into an intellectual and moral independence. The split saved us from being devoured by the single-minded fanaticism of many of our contemporaries.

My generation was the first in Mexico to live world history as its own, especially the international communist movement. Another distinctive trait of our generation: the influence of modern Spanish literature. At the end of the last century a period of splendor in Spanish letters began that culminated in the last years of the monarchy and in those of the republic and faded away in the great catastrophe of the civil war. We read enthusiastically poets and prose writers, Valle-Inclán, Jiménez, and Ortega, as well as Gómez de la Serna, García Lorca, and Guillén.[27] In the proclamation of the republic we saw the birth of a new era. Afterwards we followed the fighting for the republic as if it was our own; Alberti's visit to Mexico in 1934 further aroused our minds. For us the war in Spain was the conjunction of a Spain open to the outside with the universalism embodied in the communist movement. For the first time the Spanish tradition was not an obstacle but a path to modernity.

Our revolutionary convictions were further confirmed by another circumstance: the change in the political situation in Mexico. The ascent of Lázaro Cárdenas to power was translated into a vigorous turn to the left. The communists crossed over from the opposition to collaborate with the new government. The politics of the popular fronts, initiated during those years, justified the mutation. The more wary among us finally accepted the new line: social democrats and socialists stopped being "social-traitors" and were suddenly turned into allies in the fight against the common enemy: nazis and fascists. Cárdenas's government was notable for its generous, egalitarian verve, for its social reforms (not always apt), for its disastrous corporativism in political matters, and for its daring and nearly always faultless international policies. In cultural matters its acts were more negative. So-called "socialist education" damaged the educational system; moreover, adopted by the government, a crude, bureaucratic, demagogic art flourished. There were countless "proletarian poems" and stories and novels were paved with "progressive" commonplaces. Associations of revolutionary artists and writers, scarcely tolerated before, swelled with the arrival of new members from who knows where who soon took over the centers of official culture.

This legion of opportunists, guided and excited by intolerant doctrinaire leaders, unleashed a campaign against a group of independent writers, the so-called Contemporáneos. These writers belonged to a generation before mine; some had been my teachers, some were my friends and amongst them were several poets that I admired and admire. If I deplored the attitude of the League of Revolutionary Writers and Artists, I was more disgusted by the rhetoric of its poets and writers. From the beginning I refused to accept the jurisdiction of the Communist Party and its party leaders in questions of art and literature. I thought that true literature, whatever its themes, was naturally subversive. My opinions were scandalous but, because I was insignificant, they were treated with scorn and indifference: they came from a young nobody. However, they were not completely ignored, as I would learn a little later. During those years I began to live a conflict that would get worse and worse with time; the clash between my political ideas and my aesthetic and poetic convictions.

In 1936 I dropped my university studies and left home. I had a hard time, although not for long: the government had established secondary schools in the provinces for workers' children, and in 1937 I was offered a post in one of them. The school was in Mérida, in far-off Yucatán. I accepted straight away: I was suffocating in Mexico City. Like a seashell, the word *Yucatán* awoke echoes in my mind that were both physical and mythological: a green sea, a calcareous plain networked with underground currents like the veins of a hand, and the vast prestige of the Maya and their culture. More than distant, Yucatán was isolated, a world closed in on itself. There was no train or road; there were only two ways to reach Mérida: a weekly plane or by sea, very slowly, in a boat that left once a month and took two weeks to travel from Veracruz to the port of Progreso. The upper- and middle-class Yucatecans were not separatists, but isolationists: when they glanced abroad they did not see Mexico City but Havana and New Orleans. And the greatest difference: the dominant native element was the Mayan, descendants from the *other* ancient Mexican civilization. The real diversity of our country, hidden by the centralism inherited from the Aztecs and Castilians, was patently obvious in the land of the Maya.

I spent some months in Yucatán. Each day I lived there was a revelation, and often an enchantment. I was fascinated by the ancient civilisation as well as by the life of Mérida, half Spanish and half Indian. For the first time I was living in the tropics, not the lush green jungle but a white, dry, flat land surrounded by emptiness. Overwhelming space: time was just the blink of an eyelid. Inspired by reading T. S. Eliot, I was moved to write a poem in which the aridity of the Yucatecan plain, a parched, cruel land that seemed to be the image of what capitalism did to man and nature (capitalism for me was what made everything abstract) by draining their blood, sucking out their substance, turning them into stone and bone. I was busy with this when the school holidays hit me. I decided to make the most of this and get to know Chichén-Itza and finish my poem. I spent a week there. Sometimes alone and sometimes accompanied by a young archaeologist, I toured the ruins in a state of mind which wavered between being perplexed and feeling spellbound. It was impossible not to admire these monuments, but, at the same time, very hard to understand them. Then something happened that interrupted my holiday and changed my life.

One morning as I was walking through the Ball Game Court in whose perfect symmetry the universe seemed to rest between two parallel walls, under a diaphanous and impenetrable sky, a space where silence converses with the wind, a game field where constellations battle, altar of terrible sacrifices: on one of the reliefs that grace the sacred rectangle one can see a defeated player, on his knees, his head rolling on the ground like a sun decapitated from the heavens, while from his severed throat sprout seven jets of blood, seven rays of light, seven serpents . . . one morning, while I was strolling around the Ball Game Court, an alleged messenger from my hotel approached me and handed over a telegram that had just reached Mérida, with the request it be delivered immediately. The telegram told me to take the first available plane because I had been invited to take part in the International Congress of Anti-Fascist Writers to be held in Valencia and other cities in Spain a few days after. There was hardly time to arrange the journey. It was signed by a friend (Elena Garro).[28] The world turned upside down. Without having left the petrified time of the Maya, I felt that I was in the vital, dazzling middle of what was happening in the world. Vertiginous moment: I was planted at the point of intersection between two times and two spaces. A flash of a vision: I saw my destiny in suspension in the air of that transparent morning like a magical ball that, five hundred years before, bounced in that very enclosure, fruit of life and death in the ritual game of the ancient Mexicans.

Four or five days later I was back in Mexico City. There I learnt the reason for the telegram: the invitation had arrived on time more than a month before but the person in charge of these matters at the League of Revolutionary Writers and Artists, a Cuban writer who had been my teacher at the Faculty of Arts (Juan Marinello), had decided to send it by sea. He had thus carried out his order and made sure it would not succeed. The poet Efraín Huerta[29] found this out thanks to some tactless secretary and told Elena Garro, who sent me the telegram. On reaching Mexico City I discovered that the poet Carlos Pellicer[30] had not received his invitation either. I told him what had happened, we presented ourselves at the offices of the League of Revolutionary Writers and Artists and were given a vague explanation which we pretended to accept and all was arranged. In a few days the Mexican delegation was complete: the novelist José Mancisidor,[31] designated by the League of Revolutionary Writers and Artists, Carlos Pellicer, and myself. Why had the organizers invited two writers who did not belong to the League of Revolutionary Writers and Artists? Once in Spain, Arturo Serrano Plaja, one of those in charge of the Spanish American contingents at the Congress—the others, if my memory does not fail me, were Rafael Alberti[32] and Pablo Neruda[33]—filled in the details: they did not feel that any of the writers at the League of Revolutionary Writers and Artists really represented the Mexican literature of those days and had decided to invite a well-known poet and a young one, both friendly to the cause and both outside the party: Carlos Pellicer and myself. It was not unthinkable that they picked me: Alberti had met me on his

39

visit to Mexico in 1934; Serrano Plaja was of my age, had read my poems as I had read his, and we were united by similar ideas and curiosities. Serrano Plaja was one of my best Spanish friends; he was of a deep, religious temperament. Neruda also knew about me and years later, referring to my presence at the congress, said that he "had discovered me." In a way that was true: at the time I had sent him my first book[34] which he read and liked, and generously told me so.

IN ERUDITE DARKNESS

My experiences in Spain were varied and crucial. I cannot really detain myself here: I am not writing my memoirs. My intention in these pages is quickly to outline the main points in a political itinerary. Elsewhere I have pointed out what those inspiring days in Spain meant to me: the apprenticeship in a fraternity faced with death and defeat; the encounter with my Mediterranean roots; the realizing that our enemies are also human beings; the discovery of criticism in moral and political spheres. I discovered that the revolution is a child of criticism and that the absence of criticism had killed the revolution. But I am now telling the story of a quest, and because of that, in what follows, I will refer above all to those incidents that awoke certain doubts in me. Let me make this clear: not doubts about the justice of our cause but about the morality of methods with which we claimed to defend it. Those doubts were the beginning of my discovery of criticism, our sole moral compass in private and in public life.

Differentiating itself from ancient religious and metaphysical principles, criticism is not an absolute; on the contrary, it is the instrument to unmask false absolutes and denounce abuses... Before continuing I ought to repeat that my doubts did not blind me to the terrible grandeur of those days, a mixture of heroism and cruelty, naïveté and tragic lucidity, obtuse fanaticism and generosity. The communists were the clearest and most complete example of that duality. For them fraternity amongst the militants was the supreme value, although subordinated to discipline. Their battalions and militia were a model of organization and in their actions proved that they knew how to unite the bravest decisions with tactical intelligence. They made efficiency into their god—a god demanding the sacrifice of each one's conscience. Rarely have so many good reasons led so many virtuous souls to commit such iniquitous actions. Admirable and abominable mystery.

My first doubt began in the train that took me to Barcelona. We Mexicans and Cubans (Juan Marinello and Nicolás Guillén)[35] had arrived in Paris a day late. There we were united with a group including Pablo Neruda, Stephen Spender, the Russian writer Ilya Ehrenburg, and others. As evening fell, when we were approaching Port Bou, Pablo Neruda made a sign to Carlos Pellicer and myself. We followed him to the restaurant-car where Ehrenburg was waiting for us. We sat down at his table and in a few minutes began talking about Mexico, a country that had interested Ehrenburg since his youth. I knew this and reminded him of his famous novel *Julio Jurenito*[36] that had a scene portraying Diego Rivera. He laughed openly and told us some anecdotes about his Montparnasse years and asked us about the painter and what he was up to. They had lived together in Paris before the Revolution. Ehrenburg did not really like Diego's painting although he was amused by his personality. Pellicer answered saying that he was a close friend and spoke admiringly about Diego's collection of pre-Columbian art. He then related, with a wealth of detail, that just before setting off for Spain he had dined with Diego in his house—an unforgettable meal—and that, amongst other things, Diego had told him that Trotsky was fascinated by pre-Hispanic art. Neruda and I raised our eyebrows. But Ehrenburg did not bat an eyelid and remained silent, saying nothing. I wanted to be on hand to help and shyly said: "Yes. He once said, if I remember correctly, that he would have liked to have been an art critic..." Ehrenburg smiled briefly and nodded with his head, then made an indecipherable gesture (of curiosity, of puzzlement?). Suddenly, he mumbled in an absent voice: "Ah, Trotsky..." and turning to Pellicer asked: "What's your opinion?" There was a pause. Neruda exchanged an anguished glance with me while Pellicer said in his bass opera singer's voice: "Trotsky? He is history's greatest political agitator... after, obviously, Saint Paul." We laughed hollowly. Ehrenburg stood up and Neruda whispered in my ear: "That Catholic poet will get us shot..."

That spicy scene in the train should have prepared me for what I would later see: faced with certain topics and people the best policy is to shut one's mouth. But I did not follow this up and, without meaning it, my opinions and impressions stirred up distrust and suspicions in the pious believers, especially the members of the delegation from the League of Revolutionary Writers and Artists who arrived in Spain soon after. In nearly all the accounts published in Mexico about this event, there have been some not always innocent confusions and this delegation's participation in the Second Congress of Anti-Fascist Writers is stated as a fact. No, the only delegates were us, as I have already stated. Pellicer, Mancisidor, and myself. The League of Revolutionary Writers and Artists delegation arrived in Spain after the Congress had closed. It was composed of several artists and one writer; its somewhat vague mission was to show "Mexico's revolutionary artists' and writers' active solidarity with the Spanish people." Mancisidor and I had decided, once the Congress was over, to stay in Spain and join the League of Revolutionary Writers and Artists delegation. That was perhaps the origin of the confusion. Such suspicions caused me various problems that, luckily, I could resolve; my ill-timed opinions were private and did not threaten public safety. I was indeed the target for warnings and reprimands from some of the Communist Party stooges, as well as friendly reproaches from Mancisidor. The writer Ricardo Muñoz Suay, very young at the time, recalled later that some leader of the Alliance for Intellectuals in Valencia had warned him to be vigilant with me as I had Trotskyite tendencies. The accusation was absurd. Certainly I refused to believe that Trotsky was Hitler's agent as Moscow propaganda, repeated by communists all round the world, proclaimed; on the other hand, I thought that what mattered was to win the war and defeat the fascists. That was precisely the policy for the Communists, the Socialists, and the Republicans; the opposing thesis—upheld by many anarchists, by POUM (Working Class Party of Marxist Unification), and by the Fourth International (Trotskyite)—consisted in affirming that the only way to win the war was, at the same time, to "make the revolution." This hypothesis was doomed beforehand by reality. But in those days the slightest deviation in opinion was deemed to be "Trotskyism." Converted into a scarecrow, Trotsky's image kept the devout awake. Suspicion made them monomaniacs. I now return to my story.

In Valencia and in Madrid I was an impotent witness to the condemnation of André Gide. He was accused of being an enemy of the Spanish people, despite having declared himself from the start of the conflict as a fervent partisan of the republican cause. Due to that perverse reasoning which consists in deducing a false fact from a correct one, the rather timid criticisms that Gide had made of the Soviet regime in his *Retour de l'URSS,* turned him *ipso facto* into a traitor for the republicans. I was not alone in rebuking those attacks, although few dared to express their dissidence publicly. Amongst those who shared my feelings was a group of writers close to the magazine *Hora de España*: María Zambrano, Arturo Serrano Plaja, Ramón Gaya, Juan Gil-Albert, Antonio Sánchez Barbudo, and others.[37] They soon became my friends. I was linked to them not only by age but by literary tastes, readings in common, and our peculiar situation faced with the communists. We hesitated between fervent attachment and invincible reservations. We quickly became frank with each other: they all resented and feared the Communist Party's continual interference in their opinions and in the running of their magazine. Some of their collaborators—the most well-known cases being those of Luis Cernuda and León Felipe[38]—had even been interrogated. Writers and artists lived under the zealous gaze of commissars turned theologians.

The censors kept an eye on the writers, but the victims of the repression were the ideological adversaries. If the fight against the enemy agents was explicable and justifiable, was it right to apply the same treatment to critics and left-wing opposition, whether anarchists, socialists, or republicans? The disappearance of Andreu Nim, a POUM leader, greatly moved us. The cafés were, as always, places for gossip, but also sources of fresh news. In one of them we heard what the press did not print: a group of European socialists and British Labor Party members had visited Spain to find out, without success, where Nim had ended up. For me it was out of the question that Nim and his party were allies with Franco and agents for Hitler. A year earlier in Mexico I had met a delegation of young people from POUM: their points of view, expounded loyally by them, did not win me over, but their attitude earned my respect. I was so sure of their innocence that I would have put my hands in a fire for them. In spite of the many spies and informers, in the cafés and bars you heard, along with gossip and insinuations, spine-chilling stories about the repression. Some were obviously fantasies, but others were real, too real. Elsewhere I have referred to my sole and dramatic interview with José Bosch in Barcelona.[39] He was living clandestinely, persecuted for having taken part in the events of May that year. His end was that of many hundreds, perhaps thousands, of anti-fascists.

The outbreak of war unleashed terror on both sides. In Franco's zone the terror was, from the start, the work of those in charge and their tools, the police and the army. It was a kind of institutional violence that carried on for many years after his victory. Francoist terror was not just a weapon during the war but a policy in times of peace. Terror on the Republican side was very different. First, it was popular and chaotic: the government had fallen apart and the entities which should have kept order were impotent. People turned to the streets and took justice into their own hands. These improvised and terrible popular tribunals were as much instruments of revenge as of elimination of the Republican regime's enemies. Gloomy popular wit called these summary executions "going for a walk." The victims, real enemies or supposed ones, were taken every night from their homes by gangs of fanatics, without legal authority; they were sentenced to death in the time it took to blink and were shot in back alleys and remote places. The walk to where they were to be executed was the "going for a walk." The Republican government managed to restore order and these "walks" were suppressed in 1937. But the succeeding Republican governments, unlike the Francoists, never kept complete control of the situation, and were once again overwhelmed. Anarchist violence was replaced by the organized violence of the Communist Party and its agents, nearly all infiltrated into the Service of Military Information (SIM). Many of these agents were foreigners and all belonged to the Soviet police. Among them were Nim's assassins, as we later found out. The Republican governments, abandoned by Western democracies abroad, and at home, victims of the violent struggles between the parties making up the Popular Front, depended more and more on Soviet aid. As dependency on the USSR rose, the influence of the Spanish Communist Party increased. Protected by this situation, the Soviet police carried out a cruel policy of repression and extermination of critics and opposers to Stalin on Spanish terrain.

All this disturbed my little ideological system but did not alter my feelings for the cause of the "loyal ones," as Republicans were then called. My case is not exceptional: the clash between what we think and what we feel is common. My doubts did not touch the basis of my convictions, the revolution still seemed, despite the deviations and roundabouts of history, the sole door out of the *impasse* of our century. What could be discussed were the means and methods. As an unconscious answer to my ideological uncertainties I decided to enlist in the army as a political commissar. Alberti's wife, María Teresa León, suggested this to me. It was an aberration. I made a few moves but the manner in which I was welcomed put me off; they told me that I lacked the right skill, and above all, what was most crucial: the backing of a political party or a revolutionary organization. I was a man without a party, a mere "sympathiser." Someone high up (Julio Álvarez del Vayo) wisely told me: "You can be more useful with a typewriter than with a machine gun." I accepted his advice. I returned to Mexico, carried out various propaganda jobs in support of the Spanish Republic, and took part in the founding of *El Popular,* a newspaper that was transformed into the organ of the Mexican left. But man proposes what God stipulates. A faceless god we call destiny, history, or chance. What is his real name?

During those years the radical press in Mexico unleashed a campaign against Leon Trotsky, granted asylum in our country. Besides the communist press, the magazine *Futuro,* in which I sometimes wrote, was notable for its virulence. The editor asked me and another young writer, José Revueltas,[40] to write a leader. "I understand your reservations," he said to me, "but you have to agree, at least, that *objectively* Trotsky and his group collaborated with the Nazis. This is not just a subjective matter, although I believe that they were conscious agents of Hitler's, but historical: his attitude was useful to the enemy, and thus, in fact, it was a betrayal." His reasoning seemed to me to be deplorable sophistry. I refused to write what I was asked and left the magazine.[41] A little later, on the 23 August 1939, the German-Soviet pact was signed and on the first of September Germany invaded Poland. I felt that not only our wings but our tongues had been cut. What could we say? A few months before I had been asked to denounce Trotsky as a friend of Hitler's and now Hitler was an ally to the Soviet Union. As I read the newspaper reports about the ceremonies that followed the signing of the pact I blushed over a detail: at the official banquet Stalin stood up and offered a toast in these words: "I know about the love that the German people profess for their Führer, and so I drink to his health."

Among my friends and companions the news was at first received with disbelief; then, almost immediately, the interpretations and justifications started. A young Spanish writer,

José Herrera Petere, more naïve than the others, told us at a meeting at the publishers Séneca, managed by Bergamín: "I do not understand the motives behind this pact but I approve. I am not an intellectual but a poet. My faith is the same as a coal miner's..." At *El Popular*, once the early confusions were over, this summersault began to be justified. I talked with the editor and told him about my decision to leave the newspaper. He looked at me in surprise and said: "It's a mistake and you'll regret it. I approve of the pact and see no reason to defend the corrupt bourgeois democracies. Don't forget that they betrayed us in Munich." I accepted that what happened at Munich had been worse than an abdication but I reminded him that all the communist policies, over the recent years, had spun round the idea of a common front against fascism. And now the initiator of this policy, the Soviet government, had broken it, started war and covered all its friends and partisans with shame. I ended by telling him: "I'm off for home because I do not understand what is happening. But I will not make any public declarations or write a word against my companions." I kept my promise. More than a break, it had been a distancing: I left the newspaper and left my communist friends. The clash between what I thought and what I felt had got wider and deeper.

A few months passed. With time my bewilderment increased. After occupying Poland, the Red Army had launched itself against Finland and was about to reconquer the Baltic countries and Bessarabia. We were witnesses to the reconstruction of the old czarist empire. In a number of *Clave*, the Mexican Trotskyite magazine which I used carefully to read, an article by Leon Trotsky appeared which irritated and perplexed me. I was bothered by his unquestioning arrogance, more a despot's than a politician's, and I was staggered by the intellectual ranting it revealed. Ranting or conceit? Perhaps both things: the conceited man blinds himself. The article was a defense of Moscow's expansionist policies and could be reduced to two points. The first referred to the class nature of the Soviet Union, the sole working-class state in the world. In spite of the bureaucratic degeneration it suffered, the USSR conserved intact its social base and its means of production. For that reason, the first duty of revolutionaries was to defend it. Years earlier, in 1929, he had said that "in case of war between a bourgeois country and the USSR what would be at stake and would have to be defended is not Stalinist bureaucracy but the October Revolution." Thus, the defense of the Soviet Union was based on its social nature: it was a historically superior society to Finnish democracy or to any other capitalist democracy. The second point was deduced from the first. In a strict sense the annexation of those countries by the USSR was not an imperialist act: "In Marxist literature," wrote Trotsky, "imperialism is understood as financial capital's expansionist policy." In reality, he made clear, it was an act of self-defense. Finally, the annexation of those countries was positive, for, with or without the will of the usurping bureaucracy, the annexation could be translated as a Sovietization, that is, the imposition of a more advanced social regime, based on collective ownership of the means of production.

Trotsky's argument, although subtle, was not very different from that of the editors of *Futuro* and *El Popular*. In both cases the answer was not the result of the concrete study of the facts based on an individual's awareness; all was referred to a superior, objective agent, independent of our will: history and the laws of social development. The same idea inspired Trotsky's book on the debated theme of ends and means: *Your Morality and Ours*. I read it at that time: at first I was dazzled, halfway through I became skeptical, and I finished it tired. In that book, rich in vituperations and generalizations, you could see with greater clarity that mixture of conceit about his own ideas and the arrogant ranting, a notable defect in his powerful intelligence. In place of divine providence or any other meta-historical principle, Trotsky placed society, moved by an immanent, chimerical logic. Dialectic was the other name for that god of history, society's driving force in perpetual motion, never static, veritable holy ghost. To know its laws was to know history's direction and its plans. For Hume, the origin of religion, its root, consisted in attributing a plan to nature and its phenomena. That claim is also the root of the Leninist pseudo-religion in all its guises, including Trotsky's complicated version and Stalin's pedestrian one. In antiquity the seers interpreted the will of the gods through bird song and other signs; in the twentieth century revolutionary chiefs became interpreters of history's arcane logic. In the name of that logic and absolved by it, they committed many iniquities with the same calm conscience as that of a religious fanatic who, with his chest covered with scapulars, kills heretics and sentences pagans to death.[42]

At the end of May of that year an armed gang, under David Alfaro Siqueiros's[43] control, burst into Trotsky's house with the intention of killing him. It was as if reality had decided to refute, not with ideas but with a terrible act, his deification of history, converted into a superior logic and a manual of morality. The assault failed but the attackers kidnapped Trotsky's secretary whom they later assassinated. The raid liquidated my doubts and vacillations but left me in the dark as to which path I should take. It was impossible to continue to collaborate with the Stalinists and their friends; at the same time, what could I do? I felt intellectually and morally defenseless. I was alone. The sentimental damage was not less deep: I had to break with many dear friends. Neither did I understand the motives that had driven Siqueiros to commit that abominable act. I had known him in Spain and we soon became friends. I saw him again in Paris when he told me that he had to make a mysterious journey on a mission and I accompanied him to the train station with his wife, Juan de la Cabada, and Elena Garro. Now I think it was an alibi to get witnesses; because already then, as we learnt later, the attack was being prepared. Neither did I understand the attitude of various friends: one of them, Juan de la Cabada, helped to hide the weapons used in the attack; another, Pablo Neruda, facilitated Siquieros's entry into Chile where he sought asylum. Neither was the attitude of the Mexican government exemplary: it turned a blind eye.

Three months later, on 20 August 1940, Trotsky had his skull smashed in. The human beast's despicable logic: the assassin wounded him in his head, source of his strength. The head, place of thought, light that guided him all his life, and in the end, doomed him. An extraordinary man both in his acts and in his writings, an exemplary character who reminded me of ancient Rome's heroes, Trotsky was brave in the way he fought, upright when faced with persecutions and calumnies, and indomitable in defeat. But he never doubted his motives. He thought that his philosophy would open the world's doors; in fact, he was more and more locked up in himself. He died in a jail of concepts. That was how the cult of history's logic finished.

At the start of 1942 I met a group of intellectuals who positively influenced the development of my political thinking: Victor Serge, Benjamin Péret, the writer Jean Malaquais, Julian Gorkin, a POUM leader, and others. (I would meet Víctor

Alba a few months later.) Sometimes the Peruvian poet César Moro would join this group.[44] We would occasionally meet in Paul Rivet's flat, the anthropologist who later was director of the Musée de l'Homme in Paris. My new friends came from the oppositional left. The most notable, and the oldest, was Victor Serge. Appointed the secretary for the Third International by Lenin, he had met all the great Bolsheviks. A member of the opposition, he was exiled by Stalin in Siberia. Thanks to efforts by Gide and Malraux, the dictator agreed to change his sentence by expelling him from the Soviet Union. I believe that Stalin released an enemy but twice: one was Serge, the other Zamyatin. Serge's presence immediately attracted me. I spent hours talking with him, and have kept two of his letters. In general, with the exceptions of Péret and Moro, both poets with ideas and tastes similar to mine, the others had retained a language prickly with formulae and dry definitions from their Marxist years. Although in opposition and dissidents, psychologically and mentally they remained trapped in Marxist scolasticism. Their criticisms revealed new perspectives but their example showed me that it is not enough to change ideas: you have to change attitudes. You have to change from your roots.

Serge's human warmth, his directness and generosity, could not have been further from the pedantry of the dialecticians. A moist intelligence. In spite of his sufferings, setbacks, and long years of arid political arguments, he had managed to preserve his humanity. He owed this, without a shadow of a doubt, to his anarchistic origins; also to his wholeheartedness. I was not moved by his ideas, but by his person. I knew that my life would not be, like his, that of a professional revolutionary; I wanted to be a writer, or more exactly, a poet. But Victor Serge was for me an example of the fusion of two opposing qualities: moral and intellectual intransigence with tolerance and compassion. I learnt that politics is not only action but participation. Perhaps, I told myself, it is not a matter of changing men but of accompanying them and being one of them... The next year, 1943, I left Mexico and only returned ten years later.

THE SOLITARIES' PATH

The years I spent in the United States were poetically and vitally enriching. In return, the interchange of ideas and opinions on political matters was almost nil. But I read and continued to ponder the earlier topics. Through Serge's recommendations I became an assiduous reader of *Partisan Review*. Each month I read George Orwell's *London Letter* with renewed pleasure. An economy of language, clarity, moral audacity, and intellectual sobriety: a virile prose. Orwell had completely freed himself from the mannerisms and byzantine thought-processes of my friends, the French Marxists and ex-Marxists, if he had ever suffered them. Guided by his precise language and vivid thinking, I could at last step on solid ground. But Orwell could not help me answer certain questions that kept me awake and had more to do with political theory. Orwell was a moralist and not a philosopher. Amongst those questions, one seemed essential because my activities and the path I should follow depended on it: what was the Soviet Union's real nature? It could not be called either socialist nor capitalist: what kind of historical animal was it? I could find no answer. I now think that the answer perhaps was irrelevant. To believe that our political and moral evaluations depend on the historical nature of a particular society and not on the acts of its government and its people was to continue to be a prisoner of the circle that enclosed the Stalinists and the Trotskyites equally. It took me several years to realize that I was faced with a fallacy.

The war was reaching its end. What would happen afterwards? Would Europe's proletariat enter into the fray and fulfil Marx's prophecy as I hoped? Without a revolution in Europe, Marxism would collapse. In fact, the nucleus of its doctrine, its fundamental principle, consisted in seeing the proletariat as a universal revolutionary class destined to change history and usher in a new era. The evaporation of the historical agent of world revolution invalidates Marxism twice over, as a science of history and as a guide to action. It was natural that in 1944 many of us were asking that question. What is amazing is that, after the Second World War and despite the absence of workers' revolutions in Europe and other industrialized nations, thousands of intellectuals from all over the world should have fastened on to the chimera of world revolution, among them writers like Sartre, Moravia, and so many more who knew about Soviet reality. Much has been written about this intellectual class's aberration, but all the explanations that I have heard seem incomplete. There's a flaw, a secret fissure in the modern intellectual's awareness. Ripped out of the totality and ancient religious absolutes, we feel a nostalgia for totalities and absolutes. This perhaps explains the impulse that led them to convert to communism and defend it. It was a perverse parody of religious communion. However, how can one explain their silence faced with the lies and the crimes? Baudelaire wrote poems to Satan and spoke about his *proud* awareness of evil. His was a metaphysical evil, a vain pretense of freedom. In the case of twentieth-century intellectuals there was no rebellion or pride: there was abjection. It is hard but one must say this.

In 1944 it was still licit to hope. Many of us hoped. Meanwhile, I was present at the founding of the United Nations in San Francisco and witnessed the first skirmishes between the Western democracies and the Soviets. The cold war began. Nobody spoke of revolution, but of dividing up the world. One day the North American press published a news item that shook us all: the discovery of the Nazi concentration camps. The details were repeated and atrocious photos appeared. The news froze my bones and soul. I had been fighting Nazism since my student years at San Ildefonso and had a vague notion that concentration camps existed but I had never imagined such horror. The extermination camps opened an unexpected vista on to human nature. They exposed to my eyes the undoubted and unfathomable reality of evil.

Our century—and with ours all the centuries: our entire history—has faced us with a question that modern reason, from the eighteenth century on, has futilely tried to evade. That question is central and essential: the presence of evil among human beings. An ubiquitous presence that continues from the beginnings of the beginning and that does not depend on external circumstances but on human intimacy. Apart from the religions, who has said anything worthwhile about evil? What have philosophy and science told us? For Plato and his disciples—as well as for Saint Augustine—evil is Nothingness, the opposite of Being. But our planet is full to the brim with the works and acts of Nothingness! In the blinking of an eye Milton's devils built the wonderful edifices of Pandemonium. Can Nothingness create? Can negation make something? Criticism, that cleans minds of cobwebs and guides us to the right life, isn't it born of negation? It is difficult to answer these questions. But it is not hard to say that the shadow of evil stains and annuls the construction of all utopias. Evil is not just a metaphysical or religious notion: it is a sensuous, biological, psychological, and historical reality. Evil can be touched, evil hurts.

My life took another leap at the end of 1945: I left the United States and lived the post-war years in Paris. I could find no trace of the revolution in Europe. On the other hand, the communist empire—for that is what the union of republics founded by the Bolsheviks became—had emerged from the conflict as stronger and greater: Stalin consolidated his tyranny abroad and at home had swallowed half Europe. The Western Alliance and the Marshall Plan restrained Russia's advance in Europe; in Asia and elsewhere, the United States and its allies suffered serious setbacks, above all in China and Korea. During this period the fatal flaw in North American democracy became obvious, a defect that Tocqueville had noticed a century before: the clumsiness of its foreign policy. The exact opposite of the Roman republic, the first nation, according to Polibius, that had a real international policy.

I found France impoverished and humiliated but intellectually very alive. Having lost its ancient artistic influence, Paris had become the center of the great intellectual and political debate of those years. Communists were very powerful in the unions, in the press, and in the world of letters and the arts. Its great figures belonged to the previous generation. They were not thinking men, but poets—and poets of great talent: Aragon and Eluard, two old surrealists. The former, moreover, wrote a dazzling, sinuous prose. A serpentine temperament. Against them were ranged various dispersed groups and independent personalities like the Catholic Mauriac, a brilliant and sarcastic polemicist. Malraux had joined the Gaullists and had lost his prestige among the young intellectuals, who were more and more biased towards the communist position. The most lucid and penetrating mind was Raymond Aron's, whose writings were little read then: his time would come. There were other individualists: one of them, still very young, Albert Camus, united in his figure and in his prose two opposing distinctions: rebelliousness and the sobriety of French classicism. Jean Paulhan, another individualist, had the guts to criticize the excessive "purges" and to face up to the intimidatory policies of the communist intellectuals. A rock in that ocean of confusions: the poet René Char. Also isolated, at the centre of the fading surrealist hosts, André Breton. But the most appreciated, read, and celebrated were Sartre and his group. His prestige was immense, in and outside Europe.

From the start I felt distant from Sartre. I should slow down a while at this point because his influence was very great in Mexico, and this indirectly contributed to isolate us, myself and others with positions similar to mine. The reasons for my distancing were poetic, philosophical, and political. The first: in contrast to what happens in Heidegger, interpreter of Hölderlin and Rilke, poetry had no place in Sartre's system. In his famous essay on literature he said this clearly: poetry dilutes meanings, makes them ambiguous—to summarize, it is halfway between the word and the thing itself, it is art but not literature. Deep down, he hated art and hence his essays on or against Baudelaire and Flaubert. My other reasons are less personal, and will take longer to explain.

I was an avid reader of Ortega y Gasset, and thanks to this my surprise at Sartre's thought was less intense than that of many of his readers. There's an undoubted link between them: both derive from German philosophy and the two of them applied this philosophy with talent, in their very personal interpretations, to cultural and political themes of our time.[45] German philosophy, with the exceptions of Schopenhauer and

Nietzsche, reeks of stuffy university cloisters; Ortega y Gasset's and Sartre's of the wind in the street, of cafés and of editorial rooms. In Ortega's case German influence was more direct, and, at the same time, less overwhelming. He never dreamt of writing treatises like Sartre's. The philosophical work of the French thinker is an intelligent application of Husserl's method and an adaptation, not lacking in originality, of Heidegger's thought. A legitimate adaptation, with his own contributions. A real *stew,* in the literal and not pejorative sense of the word: a sauce made of various ingredients that enriches and gives taste to meat or fish. In Sartre's stew those ingredients, perhaps the most substantial, are of a literary and political kind. The glaring absence in Sartre's work is modern science. Burnham[46] was amazed by Trotsky's ignorance about physics, mathematics, and logic, but Sartre's lacunae were even greater. At one extreme, Sartre was an ideologue; at the other, the most valuable, a literary man. The best and most lively aspects of his work belong to the literary essay, not to philosophy.

My reservations about Sartre were to do with politics, rather than intellectual or literary matters. His specious political casuistry, more than his dense novels and ambitious philosophical treatises, led to my rejection of him. Nearly all his political essays, his theatre and fiction spin around one idea that has been the great failure of our century: the establishment of an alleged "historical logic" like a superior moral authority, independent of human will and intentions. Or put another way: acts have no value or are not measured in themselves; neither by the virtue of those who carry them out. The meter and scales are its relation with an entity that at times is called history, and at others and more frequently, revolution. Anything that promotes the revolution is a good act, including lies or the execution of hostages, said Trotsky; anything

that hinders it is evil. The superior entity is changeable because it is made of time: it is history. However, in all its
changes it is identical to itself. Each one of its movements en-
genders a negation and each negation an affirmation. The superior authority—call it revolution, historical logic, dialectics,
or the laws of social development—possesses the ubiquity of
gods: to be everywhere and at the same time to be, like them,
an unknowable reality. A reality that constantly hides itself in
its innumerable apparitions. And who could guess the meaning of each apparition? The elect: the Committee and its General Secretary. The spurious relationship between the religions
and pseudo-religious politics appears today in all its clarity.

Revolution is a god indifferent to our passions and who re-
wards or punishes with the same unpredictable infallibility
with which the Christian god in Calderón and Tirso de
Molina's plays saves or condemns sinners. Sartre repeats in a
profane and prosaic way the theological disputes of the
counter-Reformation, and its poetic version in the Spanish
theatre of the seventeenth century. But between Sartre's ab-
stractions and Christian divinity there is an enormous difference: the latter is not an impersonal logic but a person. And
more: it is the personification of universal compassion. Another crucial difference: justice in Christian divinity is unalter-
able because it is based on a code of fixed and atemporal
meanings whilst the god of history changes continuously with
circumstances and time. The same acts can lead Bukharin to
the Presidium of the Central Committee or to the firing squad.
The meaning of our acts depends on the relation between
them and the objective necessities of the revolution. The acts
are always the same but the light that illuminates them and
qualifies them changes ceaselessly. It is staggering that Sartre seriously believed that he was a libertarian philosopher; it is less

staggering that he said that man is *condemned* to be free. This idea permeates Merleau-Ponty's brilliant commentary "Humanism and Terror" on Arthur Koestler's book *Darkness at Noon*. But Merleau-Ponty had the courage and intelligence to correct himself while Sartre persisted to the end of his days: in freedom's name he disguised the crimes of the revolutionary caesars.

Sartre's arguments were not essentially different from those that I had already heard in Madrid, Mexico, and New York in the mouths of Stalinists and Trotskyites. Exactly the opposite of what I heard Breton and Camus saying. I met the former through Benjamin Péret. I have written a long essay and various articles on Breton, as well as dealing with his ideas and person in books and studies that I have dedicated to poetry.[47] Here I restrict myself to repeating that surrealism, at the time when I met Breton and his friends, had ceased to be a flame but was still smoldering. Breton wanted surrealism, a movement with which he identified not as a missionary but as its founder, to be a revolutionary way that would be located in history and in society. He sought it in communism and in the libertarian tradition, amongst Christian heretics and in literary eccentrics, in the street and in the asylums, in occultism and in magic, in this world and in others... and he didn't find it. But he was never disloyal to his quest and his sign: he always loved Lucifer, the morning star, the angel of freedom. The morality based on the chimerical "logic of history" was and remains the morality of compromise; Breton practiced exactly the contrary; a *morality of honor*. For this reason he did not make any mistakes in what really matters and did not confuse vice with virtue nor crime with innocence. His political ideas were, simultaneously, generous and vague; his libertarian passion was not exempt from setbacks and childishness;

however, in the sphere of political morality, in an inverse way to Sartre, he was literally infallible. He said *no* and he said *yes* with the same energy and when it had to be said. Time has made him right. A man of mists and flashes of lightning, he saw further than the majority of his contemporaries.

I met Camus at an event commemorating Antonio Machado at which Jean Cassou and myself spoke. María Casares read some poems admirably, and when the session ended introduced me to him. It was an effusive meeting, followed by a few more as I have written elsewhere.[48] I felt close to Camus, first, by our fidelity to Spain and its cause. Through his Spanish friends he had rediscovered the libertarian and anarchist tradition; on my side, I had also reviewed this tradition with the utmost sympathy, as I said at a political rally on 19 July 1951 in which Camus also took part.[49] I do not owe any ideas about politics or history to Camus (nor to Breton) but something more precious: having met in the loneliness of those years an attentive friend and listened to warmhearted words. I met him when he as about to publish *The Rebel,* a profound, confused book, written quickly. His thoughts on revolt are penetrating but just a start: he did not really develop his intuition. Dazzled by the very brilliance of his formulae, he was, at times, more categorical than deep. He wished to embrace many topics and ideas at once. Perhaps I am being too severe; Camus was not and did not wish to be a philosopher. He was a true writer, an admirable artist and, because of this, enamoured of form. He loved ideas in an almost Platonic sense: as forms. But living forms, inhabited by blood and passions, by the desire to embrace further forms. Ideas made from flesh and men and women's souls. Forms dreamt up and thought out by a solitary man seeking communion: a *solitary solidarity.* His philosophical and political ideas well up from a

vision that combines modern desperation with ancient stoicism. Much of what he said about revolt, solidarity, the continuous struggle of man faced with his absurdity, remains alive
and actual. Those ideas still move us because they were not
born from speculation but from a hunger that the spirit sometimes suffers when it seeks to become embodied in the world.

Breton or rebellion; Camus or revolt. As an individual I
feel closer to the former; as a social person, to the latter. My
unattainable ideal has been to be an equal among equals. The
rebel is nearly always a solitary; his archetype is Lucifer whose
sin was to prefer himself. Revolt is collective and its people are
common people. But revolt, like summer storms, quickly dis-
sipates: the very excess of its avenging fury makes it explode
and dissolve into air. In the final pages of *The Rebel,* Camus
defends moderation. In a world like ours, that has made of excess its rule and ideal, to dare to propose moderation as an answer to our evils demonstrated a great independence of mind.
A great touch was to unite moderation with revolt: moderation or measure gives a shape to revolt, informs it and makes
it permanent. For Camus moral and political health was to be
found in a return to the Mediterranean sources of our civilization, what he called *midday thinking.* The expression and the
idea so impressed me that when I read the book I wrote a few
lines. I now dare to repeat them because they are little known:

In a recent book Camus appeals for a revolt founded on Mediterranean moderation. The Greek midday is his symbol, a fixed, vi-
brating point where the contradictions that tear us apart are
reconciled: order and freedom, revolution and love. Can we face
this midday sun? Nothing is harder in a world like ours, ruled by
twin brothers: state terror and terrorist rebellion. The return to
Mediterranean moderation—if it is not a surface classicism—en-

tails the understanding of midday: moderation is tragic. More than moderation, this is a balancing of contraries and its highest form is the heroic act that says *Yes* to destiny. To glimpse the meaning of this *moderation* is to begin to recover psychic and political health. But we moderns can only see tragedy's sun from afar...[50]

Today I would add: moderation consists in accepting the relativity of values and political and historical acts, on condition that this relativity is inserted into a vision of the whole of human destiny on earth.

I met Kostas Papaioannou in 1946. He was younger than me but my intellectual debt to him is greater than our difference in age. I have tried to portray him in a few pages, and evoked his presence in a poem.[51] Given this, I will only briefly recall him. His vitality was as great as his knowledge; his intelligence, vast and deep, although its very extent led to a love for digressions that indefinitely postponed his conclusion; his cordiality, that of a stainless tablecloth, with a jug of wine, bread, and solar fruit; he was jovial and he was sarcastic; he had kept alive a dual talent to admire and become indignant. We talked a lot, in many places and for ages. In our conversations we traveled the wide avenues and sinister alleyways of history. At times we would get lost and sometimes we chatted in silence with those bodiless beings that the ancients called "the genius of place." We would also pause for a long time at a poem, a painting, a page. Kostas loved dialogue but felt a certain aversion to the act of writing, a solitary craft. Perhaps for that reason he did not leave behind him the work those of us who knew him expected. He was survived by three valuable books, it's true, but they give a weak idea of his powerful thinking. His tradition was different, a Socratic one: his true work was his conversation and the works that he stimulated in those of us who had the luck to listen to him. Another reason for gratitude: Kostas introduced me to a fellow Greek, Cornelius Castoriadis, who later became my friend and to whom we all owe invaluable clarifications in questions of philosophy and politics.

Apart from other affinities, perhaps more profound, in the world of literature, art, and philosophy, my friendship with Kostas issued from the political worries we shared. Both of us had suffered the ideological diseases of that time and our opinions placed us in what vaguely could be called the left opposition. We were both living a moral and intellectual crisis. Kostas's material situation, despite his happy indifference, was far worse than mine; he had abandoned the Greek Communist Party and lived with difficulty in France as a political refugee without papers: his country's government had taken his passport from him. He was, thus, a victim of the left and of the right, enemies united by their loathing of criticism and independence. Kostas put up with all this in his relaxed way: his love for life was stronger than his misfortunes. Soon after meeting him I realized that he was preoccupied by the same enigma that for years had been tormenting me: what was the true historical nature of the Soviet Union? Kostas was inexhaustible on this topic and his observations cleared up many mysteries for me. For example, he proved to me that Trotsky's definition that the USSR was a degenerated workers' State was an empty formula. Indeed, how could a state belong to the workers when it was not governed by workers and when in it the working class lacked the most elementary freedoms? And he added: the formula is disconcerting, but not if it comes from a man who once proposed the militarization of workers (a plan immediately provoking Lenin's opposition). Those arguments sharpened our wit and tore down many veils but would have remained mere speculations if reality, soon after, had not resolved them in a conclusive and definitive way. I am referring to the David Rousset affair, and the communist magazine *Les Lettres Françaises*. In 1947 or 1948 I read a book by David Rousset that moved me and made me think: *L'Univers concentrationnaire*. Soon after, I read another work by the same author that similarly impressed me: *Les Jours de notre mort*. As a prisoner David Rousset had experienced the Nazi concentration camps. *Les Jours de notre mort* was a terrifying testimony; *L'Univers concentrationnaire*, a profound analysis, the first that had been written, about this *other* universe that was Hitler's camps: collective extermination centers but also dehumanizing laboratories.[52] Christian hell is not of this world but in some subworld and is a place for reprobates; the concentration camp was a historical reality of this world, not a supernatural one, peopled not by sinners but by innocents. Reading Rousset's two books prompted the same sensation that I had experienced a few years before in San Francisco when I read stories in the papers about the Nazi camps: that of falling into a cold, bottomless pit. Rousset dismantled the political and psychological mechanisms of the camps, their ideological suppositions, and described their social structure. This last notion was the most disturbing: the camps were a society, the distorted mirror of ours.

Liberated, and back in Paris, Rousset participated with Jean-Paul Sartre and other personalities in the running of an ephemeral revolutionary socialist organization. He was also leader of an association of ex-Nazi-camp prisoners. I read one morning in *Le Figaro* one of his appeals to his former companions and to international pubic opinion. He and other colleagues had received numerous reports that revealed the existence of a vast network of concentration camps in the Soviet

Union. Who were the interned? Not only political opponents and "deviationists" (mostly made up of former communists), but peasants, workers, intellectuals, housewives, believers in this or that church, in short, people from every social category. Their number approached millions.

The communist press responded furiously and accused Rousset of falsification and being an agent of North American imperialism. Intellectuals were split in their opinions. Some kept silent: although they thought Rousset was right, they had no wish to offer their enemies weapons, and above all, to favor North American imperialism. At *Les Temps Modernes,* the magazine run by Sartre and Merleau-Ponty, Rousset was accused of falling into the anti-Soviet trap and using the reactionary press in his campaign. An editorial in the magazine accepted that the reported facts were true, as were, it added, the horrors of colonialism and racial discrimination in the United States, about which Rousset had said nothing. However, the crux of the question was other: whatever the distortions of the Stalinist regime, the Soviet Union was a country *towards* socialism. It was a revolution *en panne,* but it was a revolution. This position, yet again, was not very different from Trotsky's, with a fundamental difference in favor of the Russian Revolution: Trotsky had analyzed Russian reality and had concluded that it was a matter of a "degenerate workers' State"; Sartre and Merleau-Ponty stuck to asserting the revolutionary character of the Soviet State, without bothering to prove it. The curious phrase "revolution *en panne*" (broken down) reminded me of an old Mexican friend, Enrique Ramírez y Ramírez, who in a bitter argument many years ago, when we were young, hurled the following sentence at me: "The revolution is a sin but it's a sin that works."

The scandal provoked by Rousset's denunciation lasted many months. The communist press covered him with abuse; the most violent was *Les Lettres Françaises,* the magazine edited by Aragon. The dispute reached court, there was a notorious trial, a parade of witnesses, some famous, and the amazing procession of many former communists, Stalin's victims. The magazine was found to be at fault. Rousset's lawyers had adduced the Soviet Union's Code of Corrective Work that foresaw the application of sentences *by administrative decisions and without trial.* This code was the legal instrument that recruited—the exact word—those detained in concentration camps. At the time there was a belief that, unlike the Nazi camps that were purely and directly centers of collective extermination, those in the Soviet Union had an economic function. They were, as Sartre said with a metaphor of poor taste, "the Soviet Union's colonies."[53]

Faced with the enormity of these facts, unknown in our countries, especially amongst intellectuals who refused to accept certain truths, I decided to compile the most important documents—fragments of the Code, declarations from witnesses, the plaintiffs—and publish them with a short note. Elena Garro helped me compile it. But where could we publish these documents? In Spain it was impossible: Franco governed. In Mexico it was not easy: a little before I had sent a known literary supplement a statement made by André Breton where, in an aside, in two lines, he had savaged Neruda's Stalinism and that had been sufficient for the editor, Fernando Benítez, to veto the publication. I thought about *Sur.* It did not have a large circulation but it was the best literary review in our language at the time. I wrote to my friend José Bianco, he spoke to the valiant Victoria Ocampo and a little while later, in March 1951, my report appeared, with the documents and my introduction.[54] It was a public break.

I felt a kind of freedom and waited for the comments. There were few; the only answer, typically, was silence. Or in the Mexican version: they "nobodied" me. I heard later that the spoken comments had been harsh and contemptuous. Before, in Mexico, I had been viewed suspiciously, warily; since then, distrust began to turn more and more openly and intensely into hatred. Back in those days I never imagined that the vituperations would have followed me year after year, up to today. I was worried by my psychological frame of mind, or, to put it in a more antiquated and exact way: the state of my soul anguished me. I had not only lost several friends but my previous certainties. I was floating adrift. The disintoxication therapy had not completely ended; I still had a lot to learn, and more than anything, to unlearn. But I was writing, perhaps as a compensation, or to get even. Writing opened unexplored spaces for me. In brief texts in prose—poems or explosions?—I tried to grasp myself. I set sail in each word like in a nutshell. One of these texts accurately captures my state of mind. It is titled "A Poet." I underline: *a* poet, not the poet. That poet could be me but also all poets who have gone through similar bad patches in our times. That is why I later dedicated it to a couple of friends, Claude Roy and Loleh Belon, who have been torn apart and lived these anxieties. The first part of the poem alludes to a world in which the relations between men and women are at last transparent: the liberated world we dream of and demand; the second, to the reality of our century:

Music and bread, milk and wine, love and dream: free. Grand mortal embrace of enemies who love each other: every wound is a fountain. Friends sharpen their weapons, ready for the final dialogue, dialogue to the death for the rest of life. Lovers in each others arms, conjunction of stars and bodies, cross the night. Man is food for man. Knowledge is no different from dreaming and dreaming from doing. Poetry has set all poems on fire. Words are over; images are finished. The distance between names and things abolished, to name is to create, to imagine is to be born.

"For a start, grab the mattock, theorize, be punctual. Pay your price and collect your salary. In your free moments graze until you burst: there are vast meadows of newspapers. Or collapse every night on the café table, your tongue swollen with politics. Stay silent or gesticulate: it's all the same. Somewhere someone is already plotting your sentence. There is no way out that does not lead to dishonor or the gallows: your dreams are too clear, you need a strong philosophy."[55]

THE TWO FACES OF REVOLT

Two distinct movements, in continuous interpenetration, crossed the second half of the twentieth century: the cold war and the upheavals and changes on the periphery of the developed nations. The two movements are now part of the immediate past; since the fall of the Berlin wall we have entered a new period of history. I was witness to both events and in the case of changes in underdeveloped countries, a close witness. In 1952 I spent just under a year in India and Japan; I returned to Mexico at the end of 1953; in 1962 I was back in India and lived there for six years, frequently visiting, thanks to my diplomatic duties, Sri Lanka and Afghanistan. I also traveled through Nepal and South East Asia: Burma, Thailand, Singapore, and Cambodia. During that time I followed, at first hopefully and then with more and more disenchantment, the demonstrations and revolts in the misnamed Third World. My initial sympathy is easy to explain and justify. Born of the Mexican Revolution, those revolts seemed to me to confirm our movement. At that time our Revolution was viewed rather dismissively; for Marxists it was merely an episode in the universal history of class struggle, a bourgeois, democratic revolution, also nationalistic and anti-feudal; for young intellectuals a corrupt regime and an institutional lie. The criticism of the latter was just, although their perspective was wrong. The deep meaning of the revolution lay elsewhere, as I tried to explain in *The Labyrinth of Solitude*, 1950. The revolutionary movement spread in two directions: it was the meeting of Mexico with itself and that was its historical originality and its fertility; moreover, in a parallel way, it was and remains the continuation of different attempts at modernizing the country, begun at the end of the eighteenth century by Charles III and interrupted several times. The second point was what led to argument and debate.

The revolts of people on the periphery, as I pointed out in the last pages of the second edition of *The Labyrinth of Solitude*, 1959, could and should be viewed from the perspective of the Mexican Revolution's dual process. I added, somewhat impatiently: "Nobody has interrogated the blurred and formless face of the agrarian and nationalistic revolutions in Latin America and the East in order to try to understand them as they are: a universal phenomenon that requires a new interpretation." It was true, at least in part: as the configuration of the proletarian revolution faded in the developed world, another one appeared in Asia, Latin America, and Africa. Marx and his disciples had foreseen a revolution in the most advanced countries and reality had dissipated this prediction. Determined to find an explanation for the Soviet regime—

once again, the spectre of the "logic of history"—it occurred to me that even the Russian Revolution, despite its Marxist mask, was part of the great uprising of the countries on the periphery: Russia had never been entirely European. Mao's revolution, which had seduced many European intellectuals, was another of my examples, in that case with more justification.

My suppositions stemmed from an undeniable fact: the revolution foreseen by Marx had to explode inside the system formed by the advanced industrialized countries, while the new revolution arose outside that system. For Marx the terms of history's contradictions were the proletariat and the bourgeoisie, an opposition which, in turn, was the consequence of another essential contradiction between the collective nature of industrial production and the private ownership of the means of production. Instead of this chain of oppositions we were witnessing a struggle of nations not classes; the developed and the underdeveloped countries. This opposition corresponded to another that did not depend on systems of production but belonged to history and politics: imperialism and colonialism. This series of oppositions, not the ones identified by Marxism, was the cause, the origin, of the revolution of the periphery. But was this really a revolution?

It took some years to find an answer to my question. The first and most obvious difference between the classic concept of *revolution* and the uprising of the underdeveloped countries was due to the heterogeneous nature of the latter. Revolutions are social movements that put forward a universal program of changes. Revolutionary universality does not depend on a supernatural revelation but on reason. This trait differentiates the French Revolution, the first revolution of modernity, from the so-called revolutions in the United States and England, carried out in the name of specific interests and

71

principles. The twentieth century revolts in Latin America, Asia, and Africa lacked these dual characteristics, being a universal program based on the universality of reason. I thought that it was more suitable to call these movements *revolts* rather than *revolutions*. Revolt not in the sense Camus had given the word, an individual's reaction, the slave and subjugated's answer, but in the traditional and normal sense, referring always to collectivities. The protagonists of these revolts were not individuals or social classes but nations.

On reaching this position, a contradiction became evident which made understanding this phenomenon more difficult: the concept of *nation* is Western and modern, while these revolts were the uprisings of ancient peoples and cultures against the West. Thus they appeared as struggles against the West and, at the same time, they appropriated the West's political concepts: nation, democracy, socialism. This contradiction was all the more remarkable if one took note of another circumstance: the élites leading all these revolts had been educated with European methods and, often, at European universities. The contradiction, moreover, was (and is) not only political but historical and cultural: those movements glorified their traditional cultures and, simultaneously, sought at all costs to modernize their countries. Now, modernity is a Western invention. There is no other modernity than the West's: Japan's example is conclusive. Revolts contradictorily represent the resurrection of old cultures and their Westernization.

From the start third-world movements are defined by heterogeneity and contradiction. The former impeded their unification and their ability to present a common program. The lack of a program accelerated their break-up, which in turn led to narrow tribal passions and "religious fundamentalisms," to employ a faddish but useful Anglicism. Revolts showed another face: an archaic one. It was a turning back. One example among many: the fall of the Shah, a modernizing despot who did not lead Iran to democracy but to a theocratic regime. Iran's revolt, greeted enthusiastically by many European and North American intellectuals, was a step backwards. Revolts, as their name indicates, carry in their belly contrary passions and oppositions. At their most extreme, those contradictions are resolved in explosions; in their more moderate versions, as hypocritical compromises that question the coherency and even the legitimacy of the movement. Indian nationalism is an example.

India is not really a nation but a conglomeration of peoples, languages, cultures, and religions, all combined in a democratic system of government inherited from the British administration. India is one of the few countries that managed to attain independence without falling into a dictatorship. Nationalism is an idea, like democracy, that does not appear in India's history nor in its cultural traditions: it is a concept adopted by the élite from English culture. It is not exactly a doctrine but a body of vague principles and sensibilities destined to unite, from top to bottom, the different people who make up the republic. What spontaneously unites the diverse communities, from bottom to top, is religious sensibility, including in this term the castes which are primordially religious categories. Indian nationalism, in conclusion, is best viewed as a secular sensibility (its positive aspect) born of the struggle against British domination and adopted by a minority educated by the English. However, successive governments in India, since independence and without excepting the talented and civilized Nehru, have not hesitated to resort to force to repress the different nationalistic movements in the interior of the republic. Once independence was attained, Indian nationalism changed direction; it did not defend the people from foreign domination but imposed its authority upon them. Kashmir is an example, for it is not Indian culturally or historically and, above all, the majority of its population did not want it. Further cases: the Sikhs in Punjab and the Nagas in Assam.

Heterogeneity and contradiction are frequently resolved when often monstrous, even grotesque, hybrid political regimes spring up. Bizarre inventions of the pathology of history, like the controlled democracy in Indonesia or the various socialisms that flourished in some African and Asian countries. All these regimes have one thing in common: the central figure, the sun of each system, was a man who acted as guide, master, conductor, and chief. Tyrannies disguised as socialism, satrapies with the names of republics. Although many of these dictatorships have vanished, especially in Latin America, some authoritarian islands still survive, and in many places democracy faces all kinds of difficulties. Socialism for Marx and Engels would be the result of industrial development; it was scandalous that many Marxists should condone, without blinking, the farce of various governments in Asia and Africa with their insistence on converting socialism into a method of industrial and economic development. Socialism in underdeveloped countries was, from a theoretical point of view, senseless, and from a political and economic one, a colossal disaster. It led to ruin.

The most glaring example, sadly, is Castro's regime. It began as an uprising against a dictatorship as much as a reaction to the clumsy politics of the United States. It stirred up enormous sympathies all round the world, above all in Latin

America. It also awoke mine although, once bitten twice shy, I tried to keep my distance. As early as 1967, in a letter sent to Roberto Fernández Retamar,[56] prominent personality at the Casa de las Américas, I said: "I am a friend of the Cuban revolution for what it owes to Martí[57] not to Lenin." He did not answer. How could he? The Cuban regime looked more and more like Stalin's, not Lenin's, but on a smaller scale. However, many Latin-American intellectuals, obliterated by the seduction of ideology, still defend Castro in the name of the principle of non-intervention. Do they perhaps ignore the fact that this principle is based on another, the "freedom of self-determination"? A freedom that Castro, for more than thirty years, has refused the Cuban people.

The contradictions and deprivations that I have pointed out do not completely dismiss those movements. They were not revolutions but great explosions, uprisings of oppressed peoples and humiliated cultures. A kind of chain reaction, equally confused, legitimate, and necessary. It was an awakening. And what about the failures, the sacrificed lives, the lost opportunities, the mistakes and horrors, the grotesque tyrants? At times mistakes can be fruitful and losses can be warnings, lessons. Would that those nations could have learnt from their calamities. Cornered between tradition and modernity, between a living but inert past and a future reluctant to turn into a present, they have to escape the double danger that threatens them: one is petrification, the other is the loss of identity. They have to be what they are and be something else: to change and to last. To manage this they will have to find ways and aims of developing more akin to their temper. These countries, victims of delirious leaders, lack political imagination. Real imagination is born out of criticism: it is not escaping reality but confronting it. The exercise of criticism demands intelligence and likewise character, moral rigor. The criticism I am proposing is above all self-criticism. Its mission is to uproot lies, the evil that undermines the élites of those countries, especially their intellectuals, hurling them into mirages and daydreams. Without this moral reform, social and economic changes will turn into cinders... I had reached these conclusions at the end of my stay in India when in October 1968 the Mexican government's repression of the student movement forced me to quit the diplomatic service.[58]

As in 1950, my break in 1968 was spiritually healthy. It was opening a blocked door and getting out to breathe pure, sharp mountain air. I do not disown the years spent in the Mexican diplomatic service, on the contrary I recall them with gratitude. Apart from the fact that, *grosso modo,* I was nearly always in agreement with our foreign policy, I could travel, know countries and cities, deal with people of diverse trades, languages, races, capacities, and, in the end, I could write. My career, if I can call it that, was obscure and sluggish, so much so that I sometimes had the far-from-disagreeable impression that my superiors had completely forgotten about my existence. My insignificance prevented me from having the slightest influence on our foreign policy; on the other hand, it left me free. When, after twenty years service, the person who was then secretary for foreign affairs, Manuel Tello, offered me the post of ambassador, he did so in a somewhat abrupt and frank way in these words: "I can only offer you India. Perhaps you might have wanted more, but, bearing in my mind your record, I hope you'll accept it." I was not offended by his words or by the tone of his take-it-or-leave-it offer. I accepted straight away. For a start India appealed to me enormously, something that the high-ranking official could not guess; moreover, I had to grab the bull by

the horns, and India turned out to be a magnificent bull, like
Góngora's one:

Half moon the weapon on his brow
and the sun all the rays of his hair[59]

I left the post relieved, though sad to leave India. I taught in
some North American and European universities and returned
to Mexico in 1971, and that same year, thanks to the editor
of the newspaper *Excélsior,* Julio Scherer, I edited the magazine
Plural. In 1976, with little money, plenty of enthusiasm, and
several friends, I founded another magazine, *Vuelta.*[60] We con-
ceived of *Plural,* and then *Vuelta,* as primarily literary and artistic magazines open to the winds of their times, attentive to
problems and themes in the life and culture of our day, without excluding political issues. In questions of politics, our criticisms opened out in various directions: the Mexican political
system, founded on excessive presidentialism and on the hege-
mony of a party created by the state; the Soviet totalitarian system with its satellites and the Chinese one with its satellites;
the dictatorships, especially those in Latin America; the politics of the Western liberal democracies, particularly the United
States. Let me clarify once again that criticism of the capitalist
democracies has always seemed essential to me: I have never
seen them as a model. However, my enemies have not ceased
calling me "right wing" and "conservative." I am not sure today
what these antiquated adjectives mean, if they ever meant anything; nevertheless, it is not hard to guess the reason for these
slurs: since 1959 I have refused to equate liberal capitalist
democracies with totalitarian communist regimes.

Without closing one's eyes to their terrible flaws, one can
say that Western democratic societies are endowed with free

institutions. The same can be said, with obvious disclaimers, about the imperfect democracies of other areas, including the peculiar Mexican regime, on its way out. We should defend these institutions and defend the germs of freedom they contain, not annul them. That was the bias in *Plural* and today it is the one in *Vuelta*. Criticism of the Mexican system was difficult but did not stir up the debates, insults, and defamations which our denunciation of Soviet totalitarianism provoked. This is not odd: many Mexican intellectuals, for more than half a century, have suffered ideological intoxication. Some still have not been cured. The same could be said of other Latin-American countries... And here I end my recollection of that long stage that started, for me, in 1950 and that closes with the collapse of the totalitarian communist systems. What follows is the present, vast territory of unpredictability.

NIHILISM AND DEMOCRACY

Around 1980 the crisis of the Soviet empire began to be clearly apparent; it accelerated over the following years until its dissolution in December 1991. Although many of us thought that the system would soon cave in, we were all surprised by the speed of the process and the relatively pacific way it happened. It was assumed that the *nomenklatura* would defend its privileges as they had won them: with blood and fire. It was not to happen: it was demoralized. The awareness of the illegitimacy of its power must have been crushing in the last years. In matters of history all explanations are relative; with that reservation other circumstances that contributed decisively to its collapse can be cited. The first is the nature of the Russian people. To glimpse their complexity, their sudden changes, their periods of inertia followed by moments of frenzy, their illuminations and darkness, it is enough to read their great writers. The second, the economy. The "pacific competition" with the West ended in failure: it was obvious that communism would never catch up with capitalism, let alone overtake it. And the third: economic disaster combined with something more serious: the armament race with the United States left the Soviet Union breathless and, literarily, made it kiss the earth.

Very few predicted that the bankruptcy of the communist system would also be that of the Russian empire, its czarist inheritance. 1991 saw the disintegration of a political construction begun five centuries before. Forever? Nobody knows: history is a box of surprises. In any case, apart from being hypothetical, the reconstitution of the Russian empire is not a task for tomorrow. It is obvious, anyhow, that the disintegration has strengthened nationalisms. The only ideology that has survived the crises, wars, and revolutions of the nineteenth and twentieth centuries has been nationalism. The end of the cold war and the appearance of new centers of economic power, that of Japan and the Pacific rim and the European Community with Germany at the hub, as well as the possible formation of a common market in the Americas, would make the construction of an international order based on three great economic and political blocks a probability. This project now faces a formidable obstacle: the resurgence of nationalisms. Like the indetermination particle in physics, nationalism makes all political calculations unstable. It is everywhere, it dynamites all the buildings and exacerbates willpower. Some argue that the nation-state, the great political invention of modernity, has fulfilled its task and is now ineffective. Daniel Bell says that the nation-state is too small to face the great international problems and too big to solve those of the smaller nations. In short, it is reproached for neither being an empire nor a simple principality. Perhaps the solution does not lie with its disappearance but in its transformation: to convert it into an intermediary between small nationalities and blocks of nations. The concept of sovereignty would, naturally, also have to alter: today it is absolute; it must become relative.

Unfortunately, the question of nationalism is not one for political logic to solve: nationalism introduces a passionate element that is irreducible to reason and intolerant and hostile to any other point of view. More seriously, it is a contagious passion. Based on what is particular and different, it is associated with everything that divides one community from another: race, language, religion. Its alliance with religion is common and lethal for two reasons. The first because religious ties are the strongest; the second because by its nature religion, like nationalism, resists mere reason. Both are based on faith, that is, on what lies beyond reason. Thus, the resurgence of nationalisms and that of religious "fundamentalisms" faces us with a certain danger: either we can integrate them in larger units or their proliferation will lead us to political chaos, and, inevitably, to war. If the latter happens, the idea of those who view history as a senseless repetition of horrors, a monotonous succession of slaughters and empires born and dying in flames, will have been confirmed.

I am not proposing that nationalisms be eradicated. That would be impossible, and anyhow, disastrous: without them peoples and cultures would lose individuality, character: they are the lively element in history, the salt that ensures variety in every community. I have been and remain a supporter of diversity. I believe in the unique genius of each country; I also believe that the great creations, be they collective or individual, are the result of the fusion of different, even opposing el-

ements. Culture is hybridity. Empires end by becoming petrified after mechanically repeating the same formulae and mul-
tiplying images of deified Caesars. The cure for nationalism is
not an empire but a confederation of nations. The Greeks,
hypnotized by their worship of the city-state, did not or could
not transform one of their most daring political creations: the
amphictyony, a veritable confederation. They paid for such
blindness: Alexander dominated them, as Rome did later. In
the twentieth century we escaped being dominated by the
German and the Russian Empires: will we be as blind as the
Greeks and fall under the dominance of a new Rome?

Faced with Soviet expansionism, the democratic nations
followed a policy of "give and take." Flexibility can be a virtue,
as long as one does not give in to abdication and surrender.
Western government policies were never a model of coher-
ence, and were tied to unpredictable changes. Modern democracies depend on the often whimsical swings of public opinion
and are unable to formulate and successfully carry out wide-
ranging and long-term foreign policies. Instability is one of the
stigmas of modernity. However, the United States, and their
allies' policies, although firm and essentially defensive, were
successful. One of the factors of that success was the Kremlin's
cautiousness, which in the end was counter-productive for the
Russians. The reasons for such caution are numerous. In what
follows I will briefly enumerate them.

The first is to do with history. Unlike the French and the
Germans, always quick in attack, Russian policies, as much
under the Czars as under Stalin and his successors, have been
slow, calculating, and cunning. The prudence of a giant who
is unsure of the ground he is stepping on, especially if it is foreign. Let me add that these hesitations were not solely due
to the traditional psychic insecurity of the Russians, but were

determined, without a shadow of doubt, by the awareness that the Soviet leaders had of their technological and industrial inferiority. Before, economic inferiority was not a true obstacle if military superiority was taken into account, as the nomadic empires of the past knew and exploited better than anybody. The Soviet Union aspired to military superiority, and in many ways attained it, but, in the twentieth century, weapons supremacy is not sufficient; military power depends on and is subordinate to technical and industrial potency. In the final analysis: the atomic bomb. As an instrument of persuasion it was decisive and saved us from a world catastrophe. Not one of the great powers dared use it: it was a suicide weapon that would annihilate whoever launched it and the enemy. Atomic weapons brought a dismal balance into play and thus prevented the Soviet Union from possibly launching them. Had they done so they might have dominated Europe for some time. Although the liberal democracies, above all the United States, powerfully contributed to the rout of communism, the primary agent in this rout was communism itself. The Bolsheviks' and their successors' efforts to modernize their country, sacrificing democratic values, cost more blood than that of the autocrats, Peter and Catherine. The result was worse: the ruin of Russia.

Did the rout of communism mean that capitalism won? Yes, just as long as one adds that it was not a victory for justice nor of solidarity among men. The free market has shown that it is more efficient, that's all. The consequences of a state-run economy are obvious: low productivity, stagnation, improper use and squandering of human and natural resources, Pharaonic buildings (but without the beauty of Egypt), general scarcities, slavery of the workers, and a regime of privileges for the bureaucracy. There is a brutal contrast between this panorama and the one of capitalist democracies. I remember Victor Serge's amazement when, in Brussels around 1938, after his liberation, he saw the changes that had taken place in the situation of the workers: "I had to confess," he said, "that social-democracy had done it better than we had." It is undeniable that capitalism in the second half of the twentieth century is very different from the one known by Marx and the great nineteenth-century revolutionaries. Superiority of the regimes of free enterprise? Rather: democracy's superiority. Without the freedoms it grants, free enterprise would not have developed, nor with it, as antidote and corrective, worker trade-unionism and the right to strike. Without union freedom, the fate of workers would have been very different. Assuming this premise, and accepting, without haggling, the improvement of living conditions for the majority, is it licit to ask oneself: has it been enough? The answer cannot be categorical and should be nuanced.

To begin with: this well-being includes only the developed nations. It could be stated that the situation of the countries on the periphery is due to specific factors, some to the history of those countries and others, more recently, to the irresponsible policies of their governments. Maybe, but it is not all the truth. It is impossible to deny the historical responsibility of Western imperialism, from the European expansion of the sixteenth century on. More than half humanity lives on the margins of the developed world, between poverty and misery; its economic function can be reduced to providing prime materials for the industrialized countries. Furthermore, inequality appears in developed countries, although it affects but a minority of people. The situation has worsened over the last years: it is enough to travel around the great cities in the United States or Europe to realize that they are beginning to be peopled by beggars and down-and-outs like Calcutta and other cities on the periphery. The market is a mechanism that simultaneously creates zones of abundance and ones of poverty. With equal indifference it hands out consumer goods and misery.

To injustice and inequality can be added instability. Capitalist societies undergo periodic crises, financial disasters, industrial bankruptcies, the rise and fall of products and prices, sudden changes in luck amongst the owners, chronic unemployment amongst the workers. Psychological anguish, uncertainty, not knowing where we will be the next day have become second nature to us. The market promotes change and technical innovations; it is also master of waste. It makes thousands of objects, all short-lived and of poor quality; for Fourier, the ideal was to produce a limited amount of objects of insuperable quality and long-lasting, but sufficient for all.[61] The market has forced us to chuck away all that we bought yesterday, and, through advertising's ubiquitous mouth, intoxicates us with the infernal drug of novelty. Twentieth-century idolatry: the worship of new things that lasts the blinking of an eyelid. The great swindle of the market, server of nothingness, Satan's rival.

For Christian T. S. Eliot the circular process of the merely natural life was reduced to an animal trinity: *birth, copulation and death, that's all...*[62] The market simplifies this black vision: produce and consume, work and spend, *that's all...* Possessed by the lust for profit, that makes it spin and spin endlessly, it feeds off us, whether we are capitalists or workers, until we are old and infirm, to throw us like so much waste into hospitals and asylums; we are one of the millstones of its mill. The market never stops and covers the earth with gigantic pyramids of trash and scraps; it poisons rivers and lakes;

turns jungles into deserts; plunders mountaintops and the planet's innards; contaminates the air, land, and water; threatens people's lives, as well as those of animals and plants. But the market is not a natural or divine law: it is a mechanism invented by men. Like all mechanisms it is blind: it has no idea where to go, its aim is to spin endlessly. Adam Smith believed that he had located a divine intention in its circular movement, an avenging purpose. Apart from being hard to prove, the effects of the market's invisible purpose are carried out over the long term while the number of its victims increase by thousands and thousands. Moreover, Adam Smith did not take into account the inequalities of international societies made up of nations of differing economic development and varying histories and traditions. Economics tends to ignore what is specific and heterogeneous in societies and cultures.

As with nationalism, I am not proposing to abolish the market: the cure would be worse than the disease. The market is necessary; it is the heart of economic activity and one of the engines of history. The exchange of things and products is a powerful bond of union between people; it has created cultures and vehicles for ideas, people, and civilization. History is universal, thanks, among other things, to commercial exchange. At times it has been brother to war; at others, transmitter of peaceful ideas and beneficial inventions. I am not suggesting its elimination: I think that if it is an instrument it could be transformed into serving justice. The idea of the market's complete freedom is a myth. In one way or another it has been influenced in its functioning as much by state intervention as by agents of production, distribution, and consumption: businessmen, technicians, workers, traders, and consumers. We must find ways of humanizing the market: otherwise; it will devour us and devour the planet.

Neither aggressive nationalisms nor market excesses exhaust the list of evils that afflict us. We feel naturally proud of our freedoms, especially that of expression. But how today can the powerful medium of advertising be useful if it only promotes and preaches a base conformism? For Goethe, reading the newspapers was a ritual; half a century later, for Baudelaire, it was detestable, a stain to be washed clean by spiritual ablutions. We are trapped in a prison of mirrors and echoes that are the press, radio, and television that repeat, from dawn to midnight, the same images and the same formulae. The civilization of freedom has turned us into a flock of sheep. But sheep are also wolves. One of the deeply saddening traits of our society is the uniformity of awareness, tastes, and ideas, bound to the cult of an unbridled, self-centered individualism.

Profit is a god that both squashes souls like identical wafers and places them against each other with bestial ferocity. The sign stamped on every body and every soul is price. The universal question is, how much are you worth? Market laws are applied equally to political propaganda and to literature, to religious preaching and to pornography, to body beauty and to works of art. Souls and bodies, books and ideas, paintings and songs have become merchandise. Freedom and education for everybody, contrary to what the Enlightenment believed, has led us not to become familiar with Plato or Cervantes but to reading comics and best-sellers. Conformity is such that even pornography has ceased to interest the masses. Art is a commercial value: it rises and falls like shares on the stock market. I could go on about this spiritual state or better, lack of spirit, but what is the use? I am not revealing anything new; I speak of known evils. We all see the stain getting bigger, drying out brains and painting on all our faces the same grin of idiotic satisfaction.

THE SPIRAL: END AND BEGINNING

What to say about all this? First of all, say it. Yesterday we spoke of the horror we felt faced with the injustices of the totalitarian communist system; we should now view the liberal democratic societies with the same logic. Its defense, always conditional and subject to caution, should continue but transformed into a critique of its institutions, its morality and its economic, social, and political practices. In an essay "Democracy: Absolute and Relative," collected in *Ideas y costumbres*, I risked a hypothesis: perhaps one of the causes of the progressive degradation of democratic societies has been the shift from an ancient system of values based on an absolute, that is, a metahistory, to a contemporary relativism. Political democracy and a civilized co-existence among people demands a tolerance and an acceptance of values and ideas different from our own. Tolerance implies, at least in public spheres, that our religious and moral convictions are not obligatory for all but only for those who share them with us. Neither the state nor society in its entirety can be identified with this or that belief: all belong to the domain of personal conscience and awareness. Democracy is a co-existence not only of people but of ideas, religions, and philosophies. The ancient religious and philosophical absolutes have disappeared or retreated into private life in modern democracies. The result is an inner void, an absence of center and direction. To this inner void, that has converted many of our contemporaries into hollow and literally soul-less beings, should be added the evaporation of the great metahistorical projects that dazzled people from the end of the eighteenth century right up to our times. One by one all have disappeared; the last, communism, vanished leaving a pile of rubble and ash.

Democratic societies have emerged strengthened from the cold war that lasted half a century. But this victory forced them to stare at each other face to face. First of all, it should be accepted that democracy is not an absolute or a project about the future: it is a method of civilized co-existence. It is not a question of changing ourselves or of going anywhere; it asks that each one of us be able to co-exist with his neighbor, that the minority accept the will of the majority, that the majority accept the minority, and that all preserve and defend individual rights. As democracy is not perfect, we have rounded it off with a balance-of-power system, imitated from the ancients. That system, as we know, consists of a wise combination of the three modes of government in Aristotle's political philosophy: monarchy, aristocracy, and democracy. The edifice is crowned by another concept: above majorities, minorities, and individuals is the empire of the law, the same for everybody. One can live indefinitely under this system although, I repeat, it does not signal any aim for society or offer a code of metahistorical values. This system does not answer the basic questions that people have posed themselves since they have been on earth. All can be summarized in the following: What is the sense of my life and where am I going? In brief, relativism is the axis of democratic society: it assures the civilized co-existence of people, ideas, and beliefs; at the same time, at the core of relativistic societies is a hollow, a void that ceaselessly enlarges and empties souls.

The Greeks, who invented democracy, did not believe in progress. Change seemed to them an imperfection: being, supreme reality, is always identical to itself. When being changes, as in Heraclitus, it happens under the harmonic guise of repetition, that is, of the return to itself: eternal rhythm of battle that is settled in an embrace, of separation that ends in union to become again separation, and so on forever. Panic about change and movement led Plato and Aristotle to venerate the circle as image of eternal being: as it spins it continuously returns to its starting point, a movement perpetually annulling itself. How can democracy, that presupposes implicitly a static society or one endowed with a circular movement, adapt itself to modern societies that worship change? This is the question I believe that a future political science will have to address. But if I am sure of one thing it is that we are living an interregnum; we are walking across a zone whose ground is not solid: its foundations, its basis have evaporated. If we wish to climb free from the marsh and not sink into mud we should quickly work out a morality and a politics.

This is not the first time that I have alluded to the need for a political philosophy. In fact, the adjective *political* is superfluous: nearly all philosophies end up as politics. What I dream of, and what perhaps will be the task of the next generation, is the need to reaffirm Kant's tradition in one fundamental way: to trace a bridge between philosophical reflection and scientific knowledge. The only people who today are asking themselves the questions that the pre-Socratics and classical philosophers asked are the physicists, especially the cosmologists, as well as the biologists (molecular biology and neurophysics, above all). If philosophy ceases to question itself on topics like the origin and end of the universe, time, space, and similar enigmas, how can it say anything authoritatively about humans and our destiny or about the art of coexistence with our equals and with nature? If it says nothing about our origins, what can it teach us about dying? I also believe that this alleged political philosophy should choose the most immediate tradition: that of liberalism and socialism. They have generated the great dialogues of the nineteenth and twentieth centuries, and perhaps the time is ripe for a synthesis. Neither can be disowned, and are present at the birth of the Modern Age: one embodies the dream of freedom, the other that of equality. The bridge between them is brotherhood, a Christian inheritance, at least for us, born in the West. A third element, the heritage of our great poets and novelists. Nobody should dare write on philosophical and political-theoretical themes without having first read and meditated on Greek tragedy, on Shakespeare, Dante, Cervantes, Balzac, and Dostoevsky. History and politics are the domain where we choose what is unique and particular: human passions, conflicts, loves, hates, jealousies, admiration, envy, what is good and what is evil, all that makes us human. Politics is a knot linking impersonal forces—or more exactly, transpersonal—with human beings. To have forgotten this concrete person is the great sin of the nineteenth and twentieth centuries' political ideologies.

Amongst the topics that surface when we reflect a little on what is happening as the century closes there is one that deserves a long essay: the differences between modern and ancient democracies. Since its birth in Athens, democracy has been invented several times. In all its apparitions, except those of the Modern Age, it was a political regime made up of a reduced number of citizens, confined to narrow territorial limits: the city-state of antiquity, medieval communes, and Renaissance cities. In these kinds of societies, citizens knew each other. Modern democracies are huge, whether in the number of citizens or in the extension of their territory, often as large as continents. More seriously: modern democracies are formed by millions of strangers. To remedy these defects representative democracy was invented. It was a solution, but does it remain one? Before answering that question, perhaps it would be useful to take note of another great difference between ancient and modern democracies. I am referring to the ways we argue and convince in political debate. It is worth halting a while around this topic.

The basis of democracy, its very reason, is a belief in the capacity of citizens freely and responsibly to make decisions about public matters. We are dealing, let me emphasize, with a belief more than with an established principle. The same objection could be posed about other forms of government. Monarchy and aristocracy rest on similar unprovable suppositions: the monarch's and senate's capacity to govern well. There is a question of an inherent risk in all systems and forms of government. Man is a creature subject always to falling into error. For this reason, when it comes to democracies, we demand as a requisite before citizens vote that there be free, public debate. Thanks to this outdoor debate a citizen becomes acquainted with matters on which he or she should vote by weighing up the pros and cons. That way errors can be reduced. In ancient democracies the means of persuasion were direct: orators spoke to the public, displaying their reasoning and dazzling with their plans and promises. Obviously, this system did not prevent the treachery of the demagogues, nor the credulity of the citizens: nor could public debate guarantee honesty and intelligence in politicians, nor is the popular vote synonymous with wisdom. The people as rulers do not make fewer mistakes than kings or senates. Hence the need to correct flaws in democracy with remedies like the balance of powers.

There is a gulf between the ancient, popular assemblies and modern practices: the Athenians knew nothing about party bureaucracies or about the influence of the written press, radio, and all-powerful television. Public debate has turned into ceremony and showbusiness. In the United States the party conventions that elect candidates are colorful fairs that waver between a circus and a football stadium. There has always been a link between theater and politics: in both, action unfolds as representation and symbol. But today the borders between each one have been completely blurred: electoral campaigns are showbusiness. Is politics already a branch of the entertainment industry? In any case, every citizen's ability freely and rationally to choose has been seriously damaged by the media that claim to embody free speech: the press, radio, and especially television. How can we preserve freedom of speech and how prevent that freedom becoming a tool for the intellectual, moral, and political trivialization that is happening today? We can only be frank: we recognize the evil, we suffer it, but cannot envisage a remedy.

The mass (ugly word) of citizens and the transformation of public debate into showbusiness are traits that degrade modern democracies. To denounce these evils is to defend true democracy. But there is another complaint that is just as worrying. Concerning both ancient thinkers and modern ones, from Aristotle to Cicero, Locke and Montesquieu, without forgetting Machiavelli, society's political health depended on the *virtue* of its citizens. The meaning of this word *virtue* has been argued over many times, for Nietzsche's interpretation is memorable, but whatever meaning one may choose, the word *always* denotes self-mastery.[63] When virtue weakens and we are dominated by passions—nearly always by inferior ones like envy, vanity, avarice, lust, laziness—the republics die out.

When we can no longer control our appetites we are ready to be dominated by someone else. The market has undermined all the ancient beliefs—many of them, it is true, were nefarious—but only one passion has replaced them: that of buying things and consuming this or that object. Our hedonism is not a philosophy of pleasure but an abdication of free will and would have scandalized equally both the gentle Epicurus and the frantic Marquis de Sade. Hedonism is not the sin of modern democracies: its sin is conformism, the vulgarity of its passions, the uniformity of tastes, ideas, and convictions.

As *virtue* weakens, the river of blood rises. Few centuries have been as cruel as ours: the two world wars, the concentration camps, the atomic bomb, the massacres in Cambodia and other atrocities. Millions and millions killed compared with which the pyramids of skulls left by the Assyrians or Gengis Khan are child's play. However, no other civilization has concealed the idea and presence of death like ours. The omnipresence of public death and the repression of private death. In all civilizations death has been highly visible, as much in public awareness as for each individual. In some societies death has been an obsession everywhere, even in its most terrifying manifestations, and other times decked out and covered in attributes that are both laughable and gruesome. I am thinking of ancient Mexico and Tibet, and also of the Egyptians and the Celts. In other cultures, without ceasing to be a constant presence, it has not been an obsession: Greece, Rome, China. Indeed, death has been an image and a crucial reality in all societies, except in ours, because it has always been associated with a spiritual transfiguration.

The vision of death as a symbol of transmutation or freedom acquires a truly transcendental meaning in Christianity and Buddhism: it is not the opposite of life but its culmination, its fulfilment, entrance door to true life. The examples of Christianity and of Buddhism are lofty but one can find something similar in all the other philosophies and religions. Death is also a fulfilment for the Stoic philosopher, the sceptic, the Epicurean and the atheist. To die your own death has been the supreme dignity not only of saints, heroes, and wise people but for all women and men. Modern democracies offer us many things but steal from us what is most essential: they steal our own deaths, that of each one of us. The waning of virtue: weakness before the facile passions, and the concealment of death. Two faces of the same fear of life, the real one, that contains death, said the poet, like the stalk the fruit.[64]

The theme of *virtue* leads me to another. There is a moment in which a meditation on history and politics faces a phenomenon which appears in all societies and which, at the same time, pierces through them: religion. Inseparable from history, where it reveals and incarnates itself, religion spreads beyond society, outside time. One of the reasons why totalitarian ideologies are so powerfully contagious and, without a doubt, the profound cause of their collapse, is their similarity to religion. Communism showed itself in more than one way as the continuation and transfiguration of Christianity: a universal doctrine for all people, a code based on an absolute value: revolution; and, to round it off, the fusion of each part with the whole, universal communion. A sixteenth-century theologian would have viewed communism as a godless caricature of the real religion, the devil's bait. None of these values appear in modern democracies, which are secular and, thus, impartial towards religions. Modern democracy postulates a prudent neutrality in matters of faith and belief. However, it is not possible, nor is it prudent, to ignore religions and lock them away in the private domain of individual conscience. Religions include a public aspect which is essential, as can be seen in one of its most perfect expressions: the ritual of mass. Naturally I am not suggesting that religion be integrated into democracy as Rousseau, the creator of civic religion, wanted. Its separation has been an immense advance and we should never forget Socrates died after having been accused of godlessness by Athenian democracy. I underline: godlessness faced with the religion of the *polis,* a political religion. The separation of religion from politics is healthy and should persist. But religion can reveal our lack and help us rediscover and recuperate certain values.

Soon after being born we feel we are a fragment detached from something more vast and intimate. This sensation is quickly fused with another: the desire to return to that totality from which we were ripped. Philosophers, poets, theologians, and psychologists have often studied this experience. Religions have been, from the beginning, the answer to this need to participate in the whole. All religions promise us that we can return to our original home, that place where contraries cease and the self is the other and time an eternal present. Shrunk to its most simple elements—I beg forgiveness for this gross simplification—the original religious experience consists of three essential notes: the sense of a totality from which we have been cut off; in the center of this living whole, a presence (a radiant emptiness for the Buddhists) that is the heart of the universe, the spirit that guides it and gives it shape, its ultimate and absolute meaning; finally, the desire to participate in the whole and simultaneously with the creative spirit that gives it life. Participation takes place through the sacraments and good works. The doorway for Christians is death: our second birth.

The sin of political religions was to have tried to reproduce in secular terms, through simulations of religious rituals and mysteries, that yearning to participate in the whole whose supreme form is communion. The transformation of religious experience into political idolatry always ends, as we know, in vast bloody lakes. But is it not right to seek inspiration in religious experience to recover one of its purest manifestations and one not linked with any faith in particular, although appearing in all of them: veneration? There is a subtle relation between veneration and participation: veneration is already participation. We venerate the world around us, and at another level, that veneration spreads to all things and living beings, to stones and trees and animals and humans. Fraternity is an aspect of participation and both are expressions of veneration. Without veneration there can be no participation or fraternity.

A contemporary example of this dialectic between veneration, participation, and fraternity is the ecological movement. At its roots, in its very depths, ecologism is but a manifestation of an experience that pushes us to venerate the natural world, the great whole, and thus participate in and with its creations. Ecologism is no substitute for religion but its roots are religious. It expresses our thirst for totality and our yearning to participate. Certainly, there are worrying traits of this movement that recall totalitarian ideologies or makes us think of reactionary fundamentalisms. I refer above all to Gnostic and Manichean dualism, a vision that locates a perpetually fertile and beneficent power, almost a divinity, in nature; facing this, the modern, pitiless, destructive civilization. Resurrection of the myth of Gaia, our mother, and her husband, the tyrant Uranus. But Gaia, mother of Titans and Cyclops, is both creator and destroyer. The ancients protected themselves from her excesses with prayers and sacrifices; we, with science and technology. Veneration, as the ancients well knew, does not exclude healthy fear . . . Now, what I want to underline is the following: ecologism, despite its odd lapses, shows us that it is possible to recover the potential to venerate. This potential is the only one that can open the doors to fraternity with people and nature. Without fraternity, democracy gets waylaid in the nihilism of relativity, waiting room for modern societies, trap of nothingness.

With these thoughts I bring my story of a search begun in 1929 to an end. Reviewing the intervening years, I realize that this pilgrimage has brought me back to my beginning. Faced with the contemporary panorama I feel that same dissatisfaction I experienced when young confronting the modern world. I think, as I used to, that we should change it, although I no longer have the strength or youth to attempt it. Nor do I know how to go about it. Nobody does. The ancient methods were not only inefficient but foul. Does this disillusioned conclusion write off my experience and that of my generation? No: the geometric figure that symbolizes it is the spiral, a line that continuously returns to its starting point and that continuously distances itself more and more from it. The spiral never returns. We never return to the past and thus every return is a beginning. The question I asked myself at the start are the same ones I ask myself now . . . and they are different. Better: not only do I put them in a different moment but confronting them is an unknown space. At the start I asked myself: what is the sense of historical change, of the birth and collapse of nations? I did not find a reply. Perhaps there isn't one. But this absence of reply is already, as will be seen, the beginning of a reply.

Humans, inventors of ideas and artifacts, creators of poems and laws, are tragic, transient creatures: ceaseless creators of ruins. Are, then, ruins the meaning of history? If that is so, what is the meaning of ruins? Who could answer this crazy question? The god of history and the logic that rules its movements is the reason behind crimes and heroisms? That many-named god has not been seen by anybody. He is all of us: is crafted by us. History is what we make. All of us: the living and the dead. But are we responsible for what the dead did? In a certain way, yes we are: they made us and we prolong their works, the good and the bad. We are all children of Adam and Eve, the human species has the same genes from the time that they first appeared on earth. History drips with blood since Cain: are we evil itself? Or is evil outside with us as its instrument, its tool? One of the Marquis de Sade's delirious characters believed that the entire universe, from stars to people, was made up of "malevolent molecules." Absurd: neither stars nor atoms, nor plants nor animals know evil. The universe is innocent, even when it sinks a continent or explodes a galaxy. Evil is human, exclusively human. But not all is evil in humans. Evil nests in their awareness, in their freedom. In there also lies the remedy, the answer to evil. This is the sole lesson I can deduce from this long, sinuous itinerary: to fight evil is to fight ourselves. And that is the meaning of history.

MEXICO CITY
2 January 1993

Imaginary Gardens: A Memoir[65]

Dear Señora Alejandra Moreno Toscano—At the end of this letter you will find some short poems, really loose stanzas, that might have figured as inscriptions on the gates and some walls of the small garden that the city authorities, on your initiative, planned to lay on a plot of wasteland in the old part of Mixcoac. I was not born there, but when only a few months old, I was forced there from Mexico City by the disasters of the Revolution. My father joined the movement led by Zapata in the south while my mother took refuge, with me, in Mixcoac, in the house of my paternal grandfather. It was there that I lived much of my childhood and adolescence, save for a period of two years in the United States (where my father sought political asylum). This explains why, when you told me about your plan and asked for my help, I was moved to accept. However, I have just visited the noisy and desolate patch that you intended to transform into a garden and I left dismayed. My disappointment with this *terrain vague* turned into depression as I walked to the nearby rotunda with the cement statue of Manco de Celaya, General Alvaro Obregón, the revolutionary president of Mexico who lost an arm in the battle of Celaya and was later assassinated by a militant

Catholic in 1928.[66] The statue is surrounded by a tattered tribe of ash and pine trees. Although it would be a lot of work, you could perhaps partially humanize this wasteland filled with the pounding and rattling of cars. But I doubt that this future garden could become the sort of quiet and secluded spot that might evoke my poems. Besides, I confess that I do not wish to be an intruder. I do not know if I left on my own or if I was thrown out: I do know that I do not belong there. Thinking about the neighborhood that I walked through today and the one of my childhood and adolescence, I wonder: what do they have in common? I tell myself it has been worse than a destruction—it has been a degradation.

Goya Street, which is an extension of the plot you wanted to transform into a garden, used to be called Flores Street. Huge trees and severe houses—it was a bit sad. The solitude of the street was brightened by the white Teresianas College, and by the schoolgirls in their white uniforms as they came in and out of school. Women's voices and birdsong, fluttering of wings and skirts. Near the end of the street was the Gs' house, now a public office. They were family friends, and sometimes I would accompany my grandfather on his visits. The large front door would open and we entered a spacious, dark hall; we were met by a Moor with a turban and scimitar—impossible not to be reminded of Venice and Othello's followers— who held high up in his right hand a light in the shape of a torch (though the bulb was always burnt out) and led the way. I remember a corridor with flower-pots on the wall, filled with white and red flowers, possibly camellias, a floor of red brick and, separated by a small balustrade, a patio with lemon and orange trees. The mistress of the house, an old lady, accompanied by some relation, waited for us in a pale-blue room. Sometimes the conversation was interrupted by the arrival of Manuelito, the sixty-year-oldish son with a tricolour sash across his chest. He approached my grandfather deferentially and invited him to his imminent inauguration as the country's president, and asked him for advice about the composition of his future cabinet. Nobody showed the slightest surprise, and the earlier conversation soon resumed.

Flores Street was dignified without being ostentatious. The neighboring Campana Street was wide, as if proud of its elegance. It advanced with curves and meanders, not because it hesitated, or was unsure of its direction: it doubled up in order to admire itself the better. It was the best street in Mixcoac. Solid houses from the early nineteenth century. Many had full-length windows. Andalusian ironwork, white lace curtains and wooden blinds. From the street you could glimpse high-ceilinged, dark, and solitary bedrooms. Hispano-Arabic reserve: real life seethed inside the house. Strong, ochre walls, spacious and shaded gardens full of birds, pedigree dogs barking, and, above the high garden walls, the waving ocean of foliage. Blue skies, deep greens and luminous white clouds. Campana Street reached the Mixcoac river. A little stone bridge, skinny dogs and children in rags. The river was a trickle of black, stinking water. The image of drought. Only the eucalyptus trees on the banks redeemed it. Years later they filled up the river and chopped down those venerable trees.

Campana Street and the river flowed into the tram station, a characterless esplanade that was again redeemed by trees. From Tacubaya to Mixcoac the trains ran along an embankment. The two lines were bordered by two rows of ash trees, a green tunnel lit up at night by the sparks from the trolley poles. The trams were enormous, comfortable and yellow. The second-class ones smelled of vegetables and fruit; farmers brought their goods in *huacal* baskets to San Juan and La

Merced. The trams went north to Mexico City, and south to San Angel and distant Tizapán of Zapatista fame. They took fifty minutes to reach the Zócalo in the center. For the ten years that I was a student I travelled in those trams four times a day: inside I did my homework and read novels, poems, philosophical tracts, and political pamphlets. In the station there was a newsstand, a few shops and a bar. Minors were forbidden to enter. So from the door I listened to the laughter and the noise of dominoes on the tables. Nearby the snow-white bakery, and, glimpsed between the door and the counter, the Asturian baker's snow-white daughters. There were bread, apples, and cheese on a tablecloth in a meadow; nostalgia for cider, bagpipes, and drums. On the other side of the esplanade was the market building, with its din of voices and colors, a dizzying confusion of smells and sweat. Under the high plateau's great sun, men, matter, passions, and centuries foment. Then, turning the corner, ah—it's the lemon tree in snow!

Near the tram station was the boys' primary school (which is still there): a rather sad, dignified building with thick walls and huge windows. The trees had been uprooted to make room for good baseball fields. I was keen on the game and made friends with the boys there. In those days, unlike today, state schools were as prestigious as private schools and that one rivalled the French Lasalle brothers school (El Zacatito) and Williams, the English college. It is remarkable that in a relatively small area, limited today by Revolución and Insurgentes Avenues, the San Antonio Calzada and the Mixcoac Plaza, there were six schools, three for boys, three for girls: two were state, two were Catholic private, and two were lay private. Four of the schools were foreign: one was Spanish, one French, and two Anglo-American.

Towards Tacubaya, along the track some thousand meters on from the state school you reached the proud red brick villas of the Limantour, an unexpectedly English view on the Mexican plateau. These dwellings had been turned into schools: Williams for boys and Barton for girls. In Williams College I finished my primary education. The teachers were English and Mexican. They cultivated the body as a source of energy and fighting. It was an education destined to produce active, intelligent animals of prey. They worshipped manly values like tenacity, strength, loyalty, and aggression. A lot of math, geography and geometry, and of course language. They taught us to use education as a weapon, or as a prolongation of our hands. We enjoyed plenty of freedom but there was a cell for the hardened offenders, and corporal punishment was not unknown (echoes of the English system). The Williams family was Anglican, some of the teachers were possibly Catholic and the others Protestant (we never knew for sure), but what predominated was a vague deism. In El Zacatito, belief was a communal matter; in Williams, "a private opinion." The building was attractive: a ludicrous but pleasing interpretation of Tudor style. The school had football and baseball fields, freezing showers, and a debating room for the older boys. Stoicism and democracy: the jet of cold water and discussion under the water. In Williams College I was initiated (without being aware of it) into the inductive method. I learned English and a little boxing, but, above all, the art of climbing trees and the art of being alone in the fork of a tree, listening to birds.[67] Forty years later, reading *The Prelude*, I discovered that Wordsworth had had similar experiences in his childhood. Perhaps true imagination, nothing to do with fantasy, consists in seeing everyday things with the eyes of our earliest days.[68]

Beyond Williams College, and still following the tracks, you reached a strange Moorish building. The Alhambra in Mixcoac! It seemed as if it had been left there by one of the genies from an Arabian tale. That Saracen fantasy had a leafy and hilly garden. In it an amazing electric train ran through tunnels, around mountains, lakes, and cliffs. This Moorish house in Mixcoac has survived the outrages of progress, although its roofs have caved in and some of the Arabic decoration has fallen from the walls. The garden is now a supermarket. Next to the Mudejar mansion, the cave of wonders: every Thursday, a half-day at school, a cinema opened its doors and for three hours my cousins and I laughed with Delgadillo, jumped with him from skyscrapers, rode with Douglas Fairbanks, ran off with the voluptuous daughter of the sultan of Baghdad, and wept with the village orphan. The years passed and this ritual changed its day, place, and gods; I reached my fifteenth year and every Sunday "*en grande tenue de soupirant,*" as Nerval said, I arrived at the Garden Cinema, not to court a living Jenny Colon, but beautiful, impalpable ghosts.[69]

Below, along the same street, you could find the Plazuela de San Juan. Opposite each other stood a tiny eighteenth-century church and two enormous houses. Two gates, a stall, a bar, and in the *plaza,* the gigantic, inevitable ash trees. Next to them, how small the church seemed! I stared at their rough bark in amazement, and touched it with unbelieving hands: it felt like stone. They were petrified time that revived through their leaves. In this little *plaza* stood our house, with the Gómez Farías' next door. At the back of that house, among pine trees, cedars, and rose bushes there was a little monument covered with honeysuckle. It was Valentín Gómez Farías's tomb. He was a Jacobin leader, and penned the first anticlerical laws. Because of his virulent anti-clericalism, the Church hierarchy refused to bury him in the parish church's small cloister. His family decided to bury him in the garden of their house. Although all this had happened a century before, his descendants, perhaps still faithful to his memory, had not moved his remains. Rumor had it that they kept his skull in a cupboard. I visited this house many times but never discovered this hidden cupboard. The small *plaza* bordered on some yellowish, flat fields where listless cows, resigned donkeys, and wild mules took their siestas. I tried to ride one and was ignominiously thrown off and kicked. There were some deep pits: the "brickworks," excavated for earth to make adobe bricks. Inside lived cave-dwellers who terrified us. In reality, they were workers who lived deep inside the pits. Where the brick works were there is now a lovely park named after a delicate poet: Luis Urbina.[70] It was designed, I think, by a Japanese, but today it is pointlessly overcrowded with pre-Columbian reproductions—a depressing union of didactic mania with nationalistic zeal. Beyond, crossing the Insurgentes thoroughfare, the graceful San Lorenzo chapel, more fitting for sparrows than human beings, surrounded by the houses of the local artisans. Those belonging to the rocket-makers, the firework poets, stood out. I used to think of Master Pereira and his apprentices as geniuses, masters of the secret of changing fire into colors, forms, and dancing figures.

Opposite the flat fields, where the houses ended and the brick-works began, lived Ifigenia and Elodio. Their adobe house almost hung over one of the enormous pits. The floor was earth. The house was painted blue and white, and was surrounded by a fence of *magueys* and a *piru* tree that was always green, and made sounds when the wind blew. By the side, in a cramped space, swayed a field of maize. Elodio and

Ifigenia came from the lower part of the Ajusco, the great mountain that dominates the valley of Mexico. Its two volcanoes are white and blue; the Ajusco is dark, and reddish. The two old Indians were colored like their mountain; they still spoke Nahua, and their Spanish, scattered with Aztec terms, was sweet and singsong. Many years before he had been my grandfather's gardener and she had left behind her a legend as a prodigious cook. I thought of them as part of my family, and they, childless, treated me as a sort of grandchild or adopted son. Elodio had a wooden leg like a pirate from a story; he was reserved and polite, except during riotous drinking sprees, and he taught me how to shoot stones with a catapult. I fought other boys with stones in furious battles. Ifigenia was wrinkled, lively, and full of pithy sayings, an old child, with a century's wisdom. More than a grandmother, Ifigenia seemed to me like a witch from a very old story. She could cast spells and cure ailments, she told me tales, gave me amulets and scapularies, and made me chant exorcisms against devils and ghosts, illnesses and evil thoughts. In Ifigenia's house I was initiated into the mysteries of the *temascal,* the traditional Aztec bath that has something in common with a Turkish bath or the Finnish sauna. But the *temascal* was not just a hygienic practice and a bodily pleasure; it was also a ritual of communion with water, fire, and the intangible creatures engendered by steam. Ifigenia taught me how to rub myself with *zacate* grass, and with the herbs she grew. She said that the *temascal* was more like a rebirth than a bath. And it was true: after each bath I felt that I had returned from a long journey to the origins of time.

Ifigenia opened the doors for me into the Indian world that had been zealously closed by modern education. (Only years later did I discover that her name was not that of an Aztec divinity but a hapless Greek girl.) Apart from this direct contact with living Indian traditions, I also learned about their history and past. Spellbound in my grandfather's library I skimmed through amply illustrated histories of ancient Mexico. It was not long before I discovered in Mixcoac itself the subject of one of the prints illustrating my grandfather's books. During a school holiday, out on a stroll with my cousins round the outskirts of the village, we discovered a mound that could have been a tiny pyramid. We returned home excited and told the grown-ups about our find. They shook their heads mockingly; they thought it was another of my cousin's inventions (she had created a mythology about mysterious beings no larger than ants who lived in the trunk and branches of a fig tree). However, a few days later we were visited by Manuel Gamio, who was an old family friend, an archaeologist and one of the founders of modern Mexican anthropology. He listened to our tale without moving his face and that very afternoon we guided him to the place of our discovery. After seeing our mound, which was later reconstructed and identified, he explained that it was probably a shrine dedicated to Mixcoatl, the god who gave his name to our village before the Conquest. Mixcoatl is a celestial warrior god; he appears in codices with his body painted blue with white spots (the stars) and a black mask: the face of the night sky.

San Juan Street was as narrow and winding as Campana Street. San Juan Street was familiar but not banal, reserved but not sullen, modest without affectation. Like all those in Mixcoac, it was paved. Years, natural catastrophes, and municipal negligence had damaged the paving. During rainstorms the street turned into a rushing stream. In the afternoons, after school, we took our shoes off to paddle in the muddy water. In September when the rain slackened, there were numerous

puddles. I would watch the clouds sail slowly above the stag-
nant water. In the dry season the ochre earth turned to dust.
Our marbles traced a fantastic geometry over the ground and
our tops left giddy spirals.

San Juan Street ends up in the Plaza Jáuregui, the heart of
Mixcoac. As I flick through a book of prints, I see images in
front of me: the kiosk, iron benches painted green, paths
through the fields used by boys and girls after mass or fiestas
at night, the chorus of ash trees and the more intimate circle
of pines. The Municipal Palace, today the cultural center, a
sober, spacious nineteenth-century building with large bal-
conies. From there the mayor, each 16 September, would
wave the flag and cheer Hidalgo and other heroes.[71] Opposite
the Municipal Palace, there is a reddish building from the
eighteenth century. It has a noble patio, robust arcades, and a
baroque chapel. Today it is a private university; in those days
it had been divided into flats. In one of them lived my aunt
Victoria; she was almost a hundred, devout and always sigh-
ing for her Guadalajara and "those walks through the Blue
Water Park." On hearing that name I would see clouds open
and sky-blue water cascade down. Slightly hidden by trees of
the inner courtyard, white like an immense dovecote, was the
Santo Domingo Convent. It is beautiful; to look at it in the
evening soothes the mind. When the religious orders disappeared it was converted into the parish church of Mixcoac.
During the month of May, at the inner courtyard's entrance,
we waited for the girls who brought flowers to the Virgin:
spikenard, white lilies, irises. On one side of the Municipal
Palace there were several houses with severe main doors, ironwork bars and gardens. On the façade of one of them there
was a plaque which said that Lizardi had written the first
Mexican novel, *El Periquillo,* there.[72]

Outside the *plaza,* on Actipan Street, was the old estate of
El Zacatito. A large building, with a patio of heavy, rectangu-
lar columns, spacious rooms, a chapel with a choir, famous to
specialists, and the rooms of the brothers, who were all
French. On the walls, crucifixes and holy prints—*imagerie
sulpicienne.* Nevertheless, the building evoked utility more
than piety. Not grace, but practical reason. Its rational pro-
portions seemed designed not to stir up anxieties but to
confirm beliefs and convictions, but without nostalgia or in-
dulgences: it was a decidedly modern school, set to teach us
how to guide ourselves through the stormy waters of the new
twentieth century. Our textbooks were excellent but purged
of liberal heresies and clean of effeminacy and sensuality, even
of the most innocent kind. In El Zacatito were spent my first
four years of primary school. I learnt (and well) the rudiments
of grammar, arithmetic and geography, Mexican history (less
well) and religious history. I ought to say: religious history was
(is) marvellous, even in the sweetened version of brothers
Charles and Antoine. In the chapel the interminable masses
bored me. To escape the torture of enforced idleness and the
hard benches I daydreamed. Thus I discovered sin, and trem-
bled at the idea of death. In the fields I played football, I
fought and was punished (hours and hours facing a wall), and
in pranks with my friends and companions I began those first
steps along the path traveled by all men and women: the cor-
ridors of time and history. One afternoon, leaving the school
at a run, I suddenly stopped; I felt I was at the center of the
world. I raised my eyes and saw, between two clouds, an open
blue sky that was indecipherable and infinite. I did not know
what to say: I discovered enthusiasm, and, perhaps, poetry.

Please excuse this long, sentimental, and ridiculous letter.
But what has what I have just written got to do with what I

saw yesterday? Everything belongs to another world, irreme- diably alien. Mixcoac has become a word that points to a re- ality I do not recognize and which does not recognize me. I am forced to tell you, however much it pains me, that I can- not collaborate in your plan. Ask me to help you in another one, somewhere else: not in Mixcoac, and least of all that area.

Stanzas for an Imaginary Garden

The first eight lines describe a somewhat rural, provincial garden. A small enclosure with two entrances. Apart from the palm tree already there, you should plant bougainvillaeas, heliotrope, an ash and a pine. You should also install a well. This first text could be placed on one of the entrances to the little garden, either as one stanza on the lintel or on the pediment, or divided into two quartets, one on each of the doorposts:

> *Four adobe walls. Bougainvillaeas.*
> *In its quiet flames, eyes*
> *can bathe themselves. The wind passes through leaves*
> *singing praises and herbs on their knees.*
>
> *The heliotrope crosses over with purple steps,*
> *wrapped in its own aroma. There is a prophet:*
> *the ash tree—and a meditative: the pine.*
> *The garden is small, the sky infinite.*

These four lines could be placed on the other entrance, on the lintel or pediment:

Happy rectangle: some palm trees,
jade fountains; time flows,
water sings, the stone is silent, the soul,
suspended in a moment of time, is a fountain.

This text could be placed in the inside of the garden. For example, on the fountain. I imagine a wall over which a curtain of transparent water falls as you read the four lines:

Rain, dancing feet and loosened hair,
ankle bitten by lightning,
falls down accompanied by drums:
the tree opens its eyes, revives.

COLOPHON

Written after visiting the place:

Populous wasteland, a few palms,
plucked feather dusters, hammering
of motors, a prison wall,
dust and rubbish, nobody's home.

Written remembering the imaginary garden:

Green survives in my ruins:
in my eyes you look and touch yourself,
you know yourself in me and in me think yourself,
in me you survive, in me you vanish.

Afterword

The late Octavio Paz insisted that the essays included here were not memoirs; he once complained of a lack of a serious biographical tradition in Hispanic letters; at the same time, what we have is rare, a poet musing on history and politics, from a committed, public angle. In fact, it's clear that Paz never had the time (perhaps not even the inclination) to write his intellectual autobiography because he was constantly writing notes and essays and letters, so much so that Paz himself has to refer his readers to these overlapping writings where he has written something pertinent to the topic he has on hand. Hence the figure of the spiral is exact; it is the figure behind his long, oft-translated *Piedra de sol,* 1957, and catches the sinuous nature of thinking itself, always beginning again, but never the same, as time rushes on, with the act of thinking similar to the same but ever-changing Heraclitean river. The idea of translating these essays occurred to me in a hotel in London over tea with Paz, his wife Marie-José, Tony Rudolf, Daniel Weissbort, and Richard Burns. Tony's Menard Press would naturally be the publisher. Years before, in another London hotel over a whisky with Paz alone, I had been surprised at Paz's elephantine memory concerning Mexican political life. Usually our chats over the years had been literary; I was fascinated with his contacts and had long seen Paz as a guide, a *maître à penser*, who introduced me through his texts and conversations with startling new writers (I, like many, had discovered Michaux, Castoriadis, Pessoa, Villaurrutia, Cernuda, Norman O. Brown—to name a few—*through* Paz's generous enthusiasms). But that evening he let me glimpse another facet of his mental life, which was closed to a foreigner living far from Mexico like myself, and that was the detailed day-to-day interlocking with Mexican and European political issues. It is telling that Paz is drawn to poets who stretch beyond the poem, who speak from a *sagesse* about worldly and spiritual matters; the poet Paz cites most in these essays is T. S. Eliot, whose lucidity in the poems constantly kindled Paz's mind, but from often antagonistic positions and starting points. What Paz offers that is unique, then, is not culture-bound to Mexican issues and thus Mexican readers, though at times he takes a deep knowledge of Mexican history as given, but a knack to see the structure of history, the moral value of the horrors and schisms of the bloody twentieth century, pertinent to any reader. As a revolutionary who has passed through several versions of "revolution," from an early identification with his own Mexican one to an idealized Marxist one, to an intellectualized surrealist one, finally to renounce the idea of revolutionary change in itself, Paz has concentrated his witness-thinking and dialogues on the grand Marxist débâcle. That Paz focuses more on Stalin's gulags than on the holocaust is testimony to his *agon* with Marxism, often filtered through Parisian left-bank debates, as his sympathy for Albert Camus makes evident.

That a poet is asked to perform on a public platform is not new to Latin America. The great Nicaraguan-born innovator,

Rubén Darío, was famously criticized by the Uruguayan thinker José Enrique Rodó for not being the "poet of America," for not taking responsibility for his poetry. Darío, a whimsical hedonist, was stung, and tried to combine poems about his bodily pleasures, his excessive, exuberant readings and sufferings, with a more politically committed stance (it led to awful poems). Throughout the 1920s and 1930s poets debated how to be revolutionary and responsible, with crucial interventions from César Vallejo and Pablo Neruda during the Spanish Civil War; in the 1960s the Cuban Revolution generated a continental exploration of making poetry not only new but radical, changing the bourgeois reader into a guerrilla, a new man or woman in Che Guevara's slogan, as many of the poets fatally decided. Darío realized that in a continent of such divergences and injusticies and tyrannies the poet should make use of his or her advantages and prestiges, a Victor Hugo-ish involvement that later Paz couldn't be deaf to. In fact Paz's poems only appear to be free of his public and political commitment, for, as Charles Tomlinson shows in his prologue, Paz defined his task of making poetic inspiration the means to change history, which for Paz was both inside and outside him, in the language of the poem itself. To become a poet meant opening his awareness to the social tragedy, exposing the life-denying ideologies, the fossilized languages, the numbed responses. Paz took to writing essays, articles, notes, prologues, letters, interviews, often appearing on television, to intervene polemically as a poet with an ethical mission, that of defending inspiration's freedom to lift you out of yourself, to become. Paz enjoyed provoking public debates and in Mexico had many detractors, but few who could match his experience and intellectual verve. It was as if he defined the poet as someone who dared to criticize ideologues, rather than as someone who just wrote poems. *Itinerary* is Octavio Paz's reckoning as a poet, on a personal level, with evil.

The prose of *Itinerary* can also be contrasted with the poems that Paz wrote throughout his life. Although he deliberately stripped them of specific allusions to his political conflicts, we can now read them as part of the same dialogue about freedom and being. One example would be his poem "Midnight Soliloquy," written in Berkeley, California in 1944, where he attacked "God, Heaven, Friendship, Revolution or Mother Country" as "eloquent, empty bladders." Another would be *Sun Stone,* 1957, with its litany of political deaths and victims from Socrates to Brutus to Moctezuma, Robespierre, Trotsky, and Madero. And I could cite more, but what counts is how Paz worked at his political insights within his poems. Indeed, Paz cites his own prose poem "A Poet" as evidence of his involvement in the Paris of the late 1940s in *Itinerary.*

My last point concerns the way these essays assume a familiarity with Mexican history and culture; a Mexican reader who has swallowed all the dates and names at school. Paz hardly alludes to this surface procession or pageant of history, but reveals the rhythm that animates it, a quest for modernity and self, for belonging and feeling alien. In notes I have tried to outline this surface knowledge so that an Anglo-American reader can also share the pleasure of being guided to another level of understanding. Paz's own notes are signaled as such, and further notes by myself have been added to aid readers with Hispanic or continental European traditions, or with figures crucial to Paz's unfolding understanding.

My translation has tried to be faithful to Paz's particular kind of clarity: that he rarely indulges in word play, or shows off at a purely verbal level; that he is careful in his choice of words and allusions; that he suddenly drops from vast, meandering generalizations to detailed specifics; that crucial sentences have no verbs; that his punctuation is a rhythm, especially his use of semi-colons. To convey the skill with which Paz matches words with thinking I have often shifted away from Latin cognates, but have always been guided by the aim of letting Paz speak through the English.

J.W.

Notes

1 Paz published *El Laberinto de la soledad* in 1950; he expanded this first edition, with an important appendix, in 1959; the English translation by Lysander Kemp, *The Labyrinth of Solitude: Life and Thought in Mexico,* 1961, is of this second edition.

2 Paz's *La llama doble,* 1993, was translated by Helen Lane as *The Double Flame: Love and Eroticism,* 1995.

3 Mixcoac is where Paz grew up, a village swallowed into monstrous Mexico City. See the Appendix. Charles Tomlinson's poem "In a Cambridge Garden," dedicated to Paz, refers to "the monoxide monotony / That taints the trees of Mixcoac— / 'There *are* no gardens,' as you said, 'except / For those we carry with us...,' " *The Door in the Wall,* 1992, 8. The "you" of the poem is Paz.

4 Antonio Díaz Soto y Gama (1880–1967) was a brilliant, outspoken orator, an outrageous socialist and a revolutionary lawyer for Emiliano Zapata, the peasant revolutionary murdered in 1919. Díaz Soto y Gama once publicly crumpled up the Mexican flag, and called it a "rag."

5 The magazine *Contemporáneos* ran from 1928 to 1931, mingling translations from the European avant-garde with Mexican writers and artists fed up with a narrow-minded nationalism generated by the Mexican Revolution. In 1971 Paz claimed that "it opened the doors of modern poetry to me," giving him "an unforgettable jolt" when he read Blake, Saint-John Perse, T. S. Eliot, Pablo Neruda, and D. H. Lawrence for the first time.

6 Paz edited *Plural,* a lively literary magazine, for its first 58 numbers from 1971 to 1976.

7 Paz's quotation "a going-towards" is from Martin Heidegger's *Sein und Zeit*, 1927, which Paz read early in a Spanish translation, glossed by George Steiner as "a repeated conviction that the enterprise of philosophy is that of a *pilgrimage* towards" both meaning and death.

8 Juan de Jáuregui (1583–1641), who began as an anti-Góngora poet and then became an admirer, was also a painter who came from Seville; he published *Las rimas sacras y profanas*, 1618, with many translations, and is famous for his long *Orfeo* written in the 1630s but published posthumously in 1684. Paz wrote on him in *Sombras de obras*, dazzled but not moved by his poetry.

9 On Paz's writings and stay in Spain, see my book *Octavio Paz*, 1986, and the section "Spain," pp. 10–17.

10 In Vermont, Paz visited Robert Frost in June 1945 at his shack; he recorded the encounter in his essay "Visita a Robert Frost," *Las peras del olmo*, 1957.

11 Paz refers to his blue-eyed, Spanish-born mother Josefina Lozano as a "provident ant" in his autobiographical meditation-poem *Pasado en claro*, 1975: "My mother, thousand-year-old child / mother of the world, orphaned by me, / selfless, ferocious, obtuse, provident / song-bird, bitch, wild sow..."

12 Paz's *El laberinto de la soledad* opens with a chapter called "The Pachuco and Other Extremes," first published as an essay in 1949 in *Cuadernos americanos*, where Paz offers a psychological and existentialist portrait of these young, rootless Mexican rebels in street gangs.

13 Paz had been reading José Ortega y Gasset's (1883–1955) essays on the problems of "Spanishness," in *Meditaciones del Quijote*, 1914, and *España invertebrada*, 1921 (translated into English as *Invertebrate Spain*, in 1937), and especially in the magazine, the *Revista de Occidente* Ortega edited from 1923–36. See Paz's essay on Ortega in *Hombres en su siglo*, 1984.

Jorge Cuesta (1903–42) was a poet-intellectual from the *Contemporáneos* group of poets, and the patio in San Ildefonso refers to the old Jesuit college turned into the National Preparatory School where Paz studied in 1931, though Paz met Cuesta in 1935 as he recalled in his memoir/essay "Contemporáneos," 1977. Cuesta "stunned him" with his intelligence, but Paz found him too intelligent for his own good, unable to produce the work that would properly represent him.

14 Paz's note. Collected in *Primeras letras*, prologue by Enrico Mario Santí, Barcelona, 1988.

15 See Paz's extraordinary biography of this intellectual Mexican nun, *Sor Juana, o las trampas de la fe*, 1986, translated by Helen Lane as *Sor Juana, or, The Traps of Faith*, 1988. Sor Juana's (1648–94) baroque poetry has been translated into English by Luis Harss as *Sor Juana's Dream*, 1986, and by Alan Trueblood, *A Sor Juana Anthology* (with a foreword by Octavio Paz), 1988.

16 Paz refers his reader to Eric Jaufret, *Révolution et sacrifice au Mexique. Naissance d'une nation*, Paris, 1986.

17 Paz here looks behind the surface of history, as I have noted in my afterword, and assumes the reader's familiarity with the icons of the Revolution: that is, that the south, around Morelos, was dominated by Emiliano Zapata (1877?–1919) and the north by Pancho Villa (1878–1923) and all the popular associations that accompany these names. Tellingly, Paz does not even mention their names.

18 Lázaro Cárdenas (1895–1970) was the most radical of all Mexico's post-Revolutionary presidents; during his six-year presidency (1934–40) he took a stand against the Nazis and fascists, nationalized the railways (1937), the oil industry (1938), and finally carried out the 1917 Revolutionary constitution's agrarian dream by redistributing fifty per cent of cultivated land back to the peasants. Cárdenas also opened Mexico to Trotsky, and to Spain's Republican civil war exiles (Mexico refused to recognize Franco), with Paz actively involved.

19 A reader familiar with Paz's work will know that he refers to short essays collected in *Corriente alterna*, 1967, translated by Helen Lane as *Alternating Currents*, 1973.

20 Paz's Latin is a set legal phrase meaning "one witness, no witness"; that is, one witness is insufficient to establish the truth of a case.

21 Paz is quoting loosely from T. S. Eliot's *Four Quartets*; from "Burnt Norton": the *leitmotif,* "And all is always now," and the closing lines, "Quick now, here, now, always— / Ridiculous the waste sad time / Stretching before and after," and from "Little Gidding": "History is now and England," *Collected Poems 1909–1962*, 1963, pp. 194–195 & 222.

22 Quetzalcoatl, D. H. Lawrence's plumed serpent (from *Quetzal*, a rare and beautiful bird) was a god incarnate, a lawgiver, a compassionate civilizer, discoverer of maize, who turned into the

planet Venus, and who, when drunk, slept with his sister, fled his country, struggled in the underworld, and was associated with Cortés by the Aztecs, and so on, explored by Irene Nicholson in *Mexican and Central American Mythlogy,* 1967; Coatlicue was Quetzalcoatl's mother, and the Aztec mother-god and life force, with a serpent petticoat; Cortés's mistress and interpreter was called La Malinche or doña Marina, for long a symbol of betrayal.

23 Paz refers to Friedrich Nietzsche's *Die Fröhliche Wissenschaft,* 1882, and Oswald Spengler's *Der Untergang des Abendlandes,* 1918 & 1922.

24 *Sur,* edited by the rich Argentine diarist and writer Victoria Ocampo (1890–1979), was a cosmopolitan literary magazine that ran from 1931 to 1975; *Contemporáneos,* see note 5; *Cruz y Raya* was a Spanish literary magazine edited by José Bergamín from 1933–36.

25 On Jorge Cuesta, see note number 13.

26 José Bergamín (1895–1983), a radical Catholic Spanish poet, editor of *Cruz y Raya* and publisher who fought on the Republican side in the Spanish Civil War, exiled himself to Mexico and Argentina, returned to Spain in 1970 and stood as a candidate for the Republican Left in elections in 1979.

27 Paz clusters together crucial Spanish writers, from the bearded, eccentric Ramón del Valle-Inclán (1866–1936), author of novels (the best a satire on a Mexican dictator, *Tirano Banderas,* 1926) and experimental plays called *Esperpentos* (farcical satires), to the 1956 Nobel Prize winning poet Juan Ramón Jiménez (1881–1958), a dominating figure on the Spanish poetry scene, and a dedicated poet who developed a Mallarmé–style poetics called *poesía pura*; he exiled himself to Puerto Rico after the fall of the Spanish Republic in 1939, although he never wrote political poetry. The prolific Ramón Gómez de la Serna (1888–1963) was a leader of the Spanish avant-garde from his café Pombo sessions in Madrid, inventing witty aphorisms called *greguerías*. Gómez de la Serna exiled himself to Buenos Aires, where he died. Although Paz went to Spain just after Federico García Lorca's (1898–1936) murder, he was not drawn to his poetry, and has not written essays on him, while he did write brilliant essays on the more intellectual poet Jorge Guillén (1893–1984) in 1966 and 1977. Guillén collected his best work under the ever-expanding title

Cántico, 1928 and up to 1950. See Norman Thomas di Giovanni (ed.), *Cántico: A Selection,* 1965.

28 Elena Garro (1920–98) was Octavio Paz's first wife and mother of their sole child Helena Paz (1939–). She was a well-known playwright, short-story writer, and novelist. See her strange novel, *Los recuerdos del porvenir,* 1963, translated as *The Recollection of Things to Come,* 1969.

29 Efraín Huerta (1914–82) lived under the shadow of Octavio Paz, edited magazines and wrote politicized poetry. Paz's obituary was collected in *Sombras de obras,* 1983.

30 Carlos Pellicer (1899–1977), an imagistic poet who traveled the world and who also collected pre-Columbian, especially Olmec artifacts. Paz has written on him in "La poesía de Carlos Pellicer," collected in *Las peras del olmo,* 1957.

31 José Mancisidor (1894–1956), a Mexican Marxist novelist who wrote "socialist" versions of the Mexican Revolution in works such as *En la rosa de los vientos,* 1941.

32 Rafael Alberti (1901–), the last survivor of Spain's brilliant mid-twentieth-century poets known as the 1927 generation, who became politicized in the 1930s, and lived in exile in Rome and Buenos Aires while Franco controlled Spain. See Paz's memoir of the years 1930–37 when he met Alberti, overlapping with *Itinerario,* published in *Vuelta* in 1984. As Paz records, Alberti was a wonderful, if theatrical, performer of his poetry.

33 Neruda (1901–73), pen-name of Neftalí Reyes, Nobel Prize winner, was Consul General for Chile in Spain when the civil war broke out, edited a literary magazine, and had just published his quasi-surreal poems, mostly written in the Far East, called *Residencia en la tierra,* 1933 & 1935 (*Residence on Earth*). His poetry changed, thanks to the Spanish Civil War, and at that time Neruda was getting closer and closer to the Communist Party position. He would later live in Mexico, where he had a public feud with Paz that almost led to fisticuffs in 1941. Paz had met him in 1937, and after the fall-out did not speak to him until they met at the first Poetry International in London in 1967. See Paz's version in his long essay "Poesía e historia: *Laurel* y nosotros," from *Sombras de obras,* 1983.

34 Paz's first book of poems was *Luna silvestre,* 1931, a collection he refused to re-publish. These intensely lyrical and apolitical

poems, in the manner of Juan Ramón Jiménez, embodied a symbolist aesthetics of "pure poetry."

35 Juan Marinello (1898–1977), a Cuban Marxist critic, specialist on the writer and martyr José Martí, one-time rector of Havana University, poet, once president of the Cuban Communist party. He was in Spain with Paz in 1937. Nicolás Guillén (1902–89), a populist black Cuban poet who adapted Cuban *son*, became president of the Writer's Union, and has been often translated into English (see *Man-Making Words: Selected Poems of Nicolás Guillén*, translated by Robert Marquez and David McMurray, 1972).

36 Ilya Ehrenburg (1891–1967), Russian journalist and novelist, friend to many Latin Americans, including Pablo Neruda, and a crucial figure in the inter-war years in France; amazingly he survived Stalin's purges. He published over eighty novels, including *Julio Jurenito*, 1922, and *The Fall of Paris*, 1942.

37 *Hora de España*, a Republican literary magazine, was founded in January 1937 and ran until the Republic fell in 1939, with Antonio Machado as its key figure.

38 Paz's note: "The first for his elegy to García Lorca and the second for his poem *La insignia*." Lorca had been murdered in Granada in 1936. Luis Cernuda's poem "A un poeta muerto (F. G. L.)" was published in *Las nubes*, Buenos Aires, 1943. León Felipe (1884–1967) read his angry protest, and very popular poem, "La insignia" in 1937 in Valencia after the fall of Málaga during the Spanish Civil War. León Felipe later exiled himself and died in Mexico City. See Paz's poem "Carta a León Felipe" in his collection *Ladera este*, 1968, and a note written in 1938, "Saludo a León Felipe," collected in *Las peras del olmo*, 1957.

39 In notes to a poem written in Spain during the civil war called "Elegía a un compañero muerto en el frente de Aragón," collected in his *Poemas (1935–1975)*, 1979, Paz evoked his first meeting with fellow-student José Bosch in 1929, and how Bosch got him to read Kropotkin and Proudhon, how they tried to start a student strike and spent two nights in prison; after further demonstrations, Bosch was finally expelled from Mexico as a foreign agitator. After years Paz finally read that he had been killed at Aragon and wrote the poem dedicated to him. A year later, in 1938, Paz bumped into the supposedly dead Bosch in Barcelona; Bosch was now fighting for the anarchists (for POUM); Paz never

saw him again. Paz's long note abounds in details and conversations, as if Bosch embodied the passions and chaos of those years.

40 José Revueltas (1914–76), Mexican novelist, member of the Mexican Communist Party, imprisoned for his politics. Paz wrote on him in *Hombres en su siglo*, 1984, pp. 141–56. His novel *El luto humano*, 1943, has twice been translated into English, most recently as *Human Mourning*, 1989.

41 Paz's note: The director of the magazine was Vicente Lombardo Toledano, but [since he was] absent during those days, Víctor Manuel Villaseñor had temporarily taken over the running. Shortly after, Villaseñor, with Narciso Bassols and some others, published the magazine *Combate*, which stood out for its defense of the German-Soviet pact.

42 Paz's note: See my essays "Latin America and Democracy" in *Tiempo nublado*, 1983, and "The Contaminations of Contingency" in *Hombres en su siglo*. So far, not translated into English.

43 David Alfaro Siqueiros (1898–1974), one of the three great Mexican muralists, rose to become secretary of the Mexican Communist Party, fought for the Republicans in the Spanish Civil War, and had to flee Mexico after the bungled Trotsky assassination, hiding in Chile, thanks to Pablo Neruda, until 1944 when he returned home.

44 Victor Serge (1890–1947), Brussels-born political activist and novelist (of Russian parents), found himself in Russia in 1919, became adviser and friend to Lenin, then Trotsky, wrote a study of the October Revolution, was arrested several times for his comments about Stalin's methods and in 1933 sent to a camp in the Urals; liberated, he reached Mexico in 1940. Serge introduced Paz to the poet-painter Henri Michaux's work. Paz: "a discovery of capital importance for me." See Serge, *Mémoires d'un Révolutionnaire, 1901–1941*, 1978, *Carnets*, 1985, and his early account of the Moscow trials, *Seize fusillés à Moscou*, 1936.

Benjamin Péret (1899–1959), close friend to surrealist leader André Breton, co-editor of the surrealist magazine *La Révolution Surréaliste*, fought in the Spanish Civil War as an anarchist, fled to Mexico with his painter-wife Remedios Varo and lived there from 1941 to 1948. In Mexico he published his influential article on a poet's morality, *Le déshonneur des poètes*, 1945, and translated Paz's long poem *Pierre de soleil* in 1959. Paz dedicated poems to him and wrote an uncollected obituary in *Les Lettres Nouvelles,* 1959.

César Moro (1906–55), Peruvian surrealist poet and painter, real name, César Quispes Asín, joined the surrealist group in Paris in 1925, and wrote his poems in French. In 1938 he moved to Mexico, where he lived for ten years and helped André Breton and the painter Wolfgang Paalen organize the surrealist exhibition there in 1938. He has been translated into English by Philip Ward, *The Scandalous Life of César Moro,* 1976.

Víctor Alba (1916–), born in Spain, fought for the anarchists (POUM) in the Spanish Civil War; exiled in Mexico he opened an avant-garde art gallery and edited a literary magazine, *Panoramas,* and later moved to the United States.

Julián Gorkín, real name Julián Gómez, a Valencian-born Spaniard, was an early member of the Spanish Communist Party who left after being asked by Moscow to murder dictator Primo de Rivera and became a founder member of the semi-Trotskyist POUM during the Spanish Civil War. He published his memoir, *Caníbales políticos: Hitler y Stalin en España,* in 1941 in Mexico.

Jean Malaquais (1908–98), a Polish Jew who moved to France in 1926, published his war diaries as *Journal de guerre,* 1943, and *Planète sans visa,* 1947, dealing with his escape from France to New York (where he became Norman Mailer's "mentor").

45 Paz's note: See "Memento: Jean-Paul Sartre" and "José Ortega y Gasset: el cómo y el por qué" in *Hombres en su siglo,* 1984.

46 I have corrected Paz's misspelling of James Burnham (1905–87), author of *The Managerial Revolution,* 1941, and *The Defeat of Communism,* 1950.

47 Paz's note: See "El surrealismo" in *Las peras del olmo,* 1957; "Constelaciones: Breton y Miró" in *Hombres en su siglo,* 1984 and "Poemas mudos y objectos parlantes (André Breton)" in *Convergencias,* 1991. For more on this background, see chapter two, "The Surrealist Years," in my study *Octavio Paz,* 1986.

48 Paz's note: See "Inicuas simetrías" in *Hombres en su siglo,* 1984. Paz visited Antonio Machado (1875–1939) in Valencia during the Spanish Civil War, and wrote about it later in 1951, collected in *Las peras del olmo,* 1957. Machado died fleeing Barcelona after its fall to Franco in 1939, by then having become the intellectual figurehead of the Republican movement; a great poet-thinker.

49 Paz's note: See "Aniversario español" in *El ogro filantrópico,* 1979.

50 Paz's note: These lines appeared in the first edition of *El arco y la lira*, 1956. I had to suppress them in the second edition for reasons of space and style: they were a digression. I am glad to be able to restore them.

51 Paz's note. See "Kostas Papaioannou (1925–1981)" in *Hombres en su siglo*, 1984. Paz also wrote a long poem, "Kostas Papaioannou (1925–1981)," where he recalls being a thirty-year-old, meeting the younger Papaioannou ("a universal Greek from Paris") in 1946 in a Parisian café and talking of Zapata, published in *Arbol adentro*, 1987.

52 Rousset's (1912–97) *L'Univers concentrationnaire* appeared in 1946, *Les jours de notre mort* in 1947. *La Société éclatée* appeared in 1973. Paz cannot have known about Primo Levi's *Se questo è un uomo*, 1947, or Robert Antelme's *L'Espèce humaine*, translated into English in 1992. For background to Rousset, see Herbert Lottman, *The Left Bank: Writers, Artists and Politics from the Popular Front to the Cold War*, Heinemann, London, 1982.

53 For background to this period, see Tony Judt, *Past Imperfect: French Intellectuals, 1944–1956*, 1992, pp. 113–15, which outlines Rousset's appeal to enquire into Soviet labor camps, published on 12 November 1950 in *Le Figaro littéraire*, as well as the trial Paz refers to, and the Parisian left's responses.

54 Paz's note. See "Los campos de concentración soviéticos" in *El ogro filantrópico*, 1979.

55 This text first appeared in Paz's *¿Aguila o sol?*, 1951. As I showed in my critical study, *Octavio Paz*, 1986 (see pp. 67–8), this prosepoem, written in Paris, opposes a surrealist libertarian poetics to a Marxist one that leads poets to Stalin's gulags and ethical dishonor (referring to Benjamin Péret's *Le déshonneur des poètes*, 1945).

56 Roberto Fernández Retamar (1930–), Cuban poet and critic, wrote an influential essay on the role of the intellectual in Latin America titled *Calibán. Apuntes sobre la cultura en nuestra América*, 1971, and translated as *Caliban and Other Essays*, 1989. He directs the prestigious cultural center Casa de las Américas in Havana, running the literary magazine of the same name.

57 José Martí (1853–95), Cuban-born, spent his life, and lost it, fighting for independence from Spain; often exiled, he wrote polemical journalism, was a poet and through his sacrifice became the figurehead for Cuban liberation, taken up by Castro and revered by all Cubans.

58 The massacre at Tlatelolco on 2 October 1968, just before the Mexican Olympic Games, which saw 350-odd students slaughtered in the Plaza de las Tres Culturas (uniting the Aztec past with a chapel from colonial times and modern tower blocks), was commented on by Paz in a poem "La limpidez," and then in a long essay, *Posdata*, 1970 (*The Other Mexico: Critique of the Pyramid*, 1972). The critic and writer Elena Poniatowska taped survivors' responses in her denunciatory book *La noche de Tlatelolco*, 1971, translated as *Massacre in Mexico*, 1975, with a prologue by Paz. The massacre was a watershed in Mexican politics as it revealed the flaws in the revolutionary ambitions of the dominant party in constant power, the PRI.

59 Luis de Góngora y Argote was Spain's greatest baroque poet, born in Córdoba in 1561 and dying there in 1627, having lived as a minor cleric. His ornate, hermetic imagery, best seen in his *Soledades*, 1613, was vilified in his life, but was rediscovered in 1927 by Lorca and his friends. Paz cites from the *Soledad primera*, lines 3 and 4; see R. O. Jones, *Poems of Góngora*, Cambridge University Press, 1966, p. 40.

60 Paz's note: See "Vuelta" in *El ogro filantrópico*, 1979. Paz continued to edit *Vuelta*, adding a publishing house of the same name, until his death in March 1998; it ceased publication in August 1998.

61 On Charles Fourier, the French utopian anarchist, see Paz's essays *El ogro filantrópico: historia y política 1971–1978*, 1979, where he refers to Fourier in "Por qué Fourier" (pp. 208–11) as more crucial than Marx; Fourier will be the "touchstone of the twentieth century."

62 The lines quoted by Paz are from T. S. Eliot's "Fragment of an Agon" in the *Collected Poems, 1909–1962*, 1974, p. 131, though he slightly misquotes. It should read: "Birth, and copulation, and death, / That's all..."

63 Paz could be referring to Nietzsche's *Thus Spoke Zarathustra* and the section "Of the virtuous" that argues that virtue is a front for laziness, hatred, revenge, righteousness, etc., rather than the self in action.

64 Paz does not name this poet, but it is Rainer Maria Rilke, who in the sixth elegy, referring to a fig tree, first develops the analogy between a ripe fruit and a natural death and who closes his tenth elegy of the *Duino Elegies* with the notion of a fruit-falling: "... the emotion / that almost bewilders us / when a happy thing

falls" (from the new complete translation by Patrick Bridgwater, 1999).

65 This piece first appeared in *The Times Literary Supplement,* July 14–20, 1989, translated by myself.

66 *Manco* means one-armed; it was applied as a nickname to Cervantes when he lost an arm in a sea-battle ("El manco de Lepanto"); the battle of Celaya, a town 150-odd miles north of Mexico City, was where Obregón, in defensive trenches, decimated Pancho Villa's attacking army of the north in 1915. Obrégon was elected president in 1920.

67 See Paz's prose poem "Jardín con niño" ["Garden with Child"] from *¿Aguila o sol?,* 1951 *(Eagle or Sun,* 1970), which has the child up in a tree, innocent about his future, seen by the poet, curved over his desk, writing laboriously his "goodbyes on the edge of the precipice." Paz's Mixcoac childhood climbing of trees returns in the same collection with the prose poem "La higuera" ("The Fig Tree"), where the tree talks back and promises him his future life.

68 Paz's reading of Wordsworth surfaced in his great autobiographic poem *Pasado en claro,* 1975, as epigraph, and literary echo.

69 From the opening lines of Gérard de Nerval's (1808–55) *Sylvie, souvenirs du Valois,* "Je sortais d'un théâtre où tous les soirs je paraissais aux avant-scènes en grande tenue de soupirant...," in *Oeuvres,* vol. 1. p. 589.

70 Luis Urbina's (1868–1934) poetry was compared to impressionistic painting by Paz in an essay of 1950; in 1942 he found Urbina lazily sensuous, rich in nuances, adding "he is not our best poet, but is one of our most loved ones," in early essays collected in *Las peras del olmo,* 1957.

71 In 1810, Father Miguel Hidalgo started the rebellion against Spain, from his village of Dolores, with his famous *grito* (shout) *de Dolores,* celebrated every 16 September as Mexico's national day. Hidalgo fought for racial equality, and land reform, but was sentenced to death by the Spaniards in 1811. Mexico was finally independent from Spain in 1820.

72 José Joaquín Fernández de Lizardi (1776–1827) wrote his picaresque novel *El periquillo sarniento* in 1816 (translated in 1942 as *The Itching Parrot*). Publishing novels in the New World colonies had been banned by the Spanish crown, and Lizardi's can be considered the first novel published in Latin America.

If you enjoyed *Itinerary*, look for
these other titles by the Nobel Prize winner

OCTAVIO PAZ

In Light of India
0-15-100222-3 (hc)
$22.00 / Higher in Canada

0-15-600578-6 (pb)
$13.00 / Higher in Canada

An Erotic Beyond: Sade
0-15-100352-1 (hc)
$18.00 / Higher in Canada

The Double Flame
0-15-600365-1 (pb)
$15.00 / Higher in Canada

The Other Voice
0-15-670455-2 (pb)
$12.00 / Higher in Canada

Printed in the United States
90990LV00002B/58/A